Atmosphere, Climate, and Change

Hurricane Andrew
GOES 7 19:00 UTC
August 25, 1992
Red: Visible
Green: Visible
Blue: Inverted 11 μm Infrared

NASA Goddard Laboratory for Atmospheres   Hasler, Pierce, Palaniappan, Manyin

Hurricane Andrew nears the east coast of the United States in this August 25, 1992 image from the GOES satellite. It is conjectured that one of the consequences of global warming could be an increase in the frequency of hurricanes and tornadoes.

Atmosphere, Climate, and Change

Thomas E. Graedel and Paul J. Crutzen

**SCIENTIFIC
AMERICAN
LIBRARY**

A division of HPHLP

New York

Library of Congress Cataloging-in-Publication Data

Graedel, T. E.
    Atmosphere, climate, and change / Thomas E. Graedel and Paul J. Crutzen.
        p.   cm.
    Includes bibliographical references and index.
    ISBN 0-7167-5049-X
    1. Climatology.   2. Atmospheric chemistry.   I. Crutzen, Paul J.,
    1933-   .   II. Title
    QC981.G63   1995
    551.6—dc20   94-23797
                                                                    CIP

ISSN 140-3213-5026-0

Printed in the United States of America.

Scientific American Library
A division of HPHLP
New York

Distributed by W. H. Freeman and Company.
41 Madison Avenue, New York, NY 10010
20 Beaumont Street, Oxford OXI 2NQ, England

1 2 3 4 5 6 7 8 9 0 KP 9 9 8 7 6 5 4

This book is number 55 of a series.

Contents

Preface

When the two of us were college students in the 1960s, almost nothing was known of the chemistry of the atmosphere and the resulting effects on the climate of our planet. It was our great good fortune to help launch these areas as fields of study and then to participate in the explosive growth of understanding as increasing numbers of outstanding scientists addressed these topics. The excitement of this work has perhaps been heightened because every living thing on Earth is affected every day by atmospheric and other planetary processes—heating by the Sun, the formation of clouds, cycles of volcanic emission and deposition—and that appreciating how these processes work is not science removed to an ivory tower but science thrusting itself into all our lives. There is perhaps no greater thrill, at least not for us, than learning how our planet goes about its business, and each day we learn more.

In studies of environmental chemistry and climate, it is quite natural for the atmosphere to take a central role, but more and more specialists in those fields find themselves entering into collaborations with oceanographers, soil scientists, hydrologists, volcanologists, plant pathologists, and hosts of other kinds of experts, because Earth's complex processes—especially atmospheric ones— very quickly override interdisciplinary boundaries. Thus, we cannot write meaningfully about the atmosphere without discussing the entire Earth system to some degree. Such an endeavor can be a wonderful intellectual exercise for authors and readers if done properly, chaotic if not. We hope we have struck the right balance in dealing with these topics. Another issue we wrestled with was whether and when to use chemical equations, cognizant of Stephen Hawking's rule that each equation included in a book would halve the sales. Our eventual decision was a compromise: although some explicit chemistry was necessary for optimum understanding, we did confine its use to Chapter 3 (and a bit in Chapter 5). Readers who wish a general picture of atmospheric change can acquire it while skimming that material, but we encourage you to spend a little time with it; the result will be a much better understanding of how the atmosphere works.

Notwithstanding our enthusiasm for learning about what we might call "the atmosphere as it functions today," it is obvious that *changes* in the atmosphere and climate have become such a media topic that concerns about the atmosphere are now a common part of our lives. Every month brings, it seems, a new potential horror story about a different threat to the planet, with the atmosphere at the forefront of many of them: acid rain, smog, the ozone hole, global warming. Thoughtful people are led to ask a few reasonable questions: Are there really so many problems? Why don't scientists understand them better? How accurately can scientists predict the future? Must people change their way of life in order to help solve these problems?

In this book, we try to answer some of these questions and to indicate why answers to some of the others still elude us. In many ways, atmospheric and climatic change are symptomatic of many topics currently under discussion in our technological society, where attempts to solve scientifically related problems often seem to be motivated more by corporate public relations or political propaganda rather than by facts or, at least by informed opinion. Practical responses to global change and steps toward the sustainable development of the planet will not be made by scientists alone. Rather, scientists must share their knowledge with society as a whole in the hope that political leaders and citizens of all kinds can work together to make rational decisions about the future of "Spaceship Earth." It is disturbing but true that many of the technological problems facing humans today have three attributes that render them very different from traditional social problems: they have developed (and can only be remedied) over a long time scale, decades to centuries or more; many of the details of their causes and consequences are likely to be uncertain at the time decisions must be made; and the results of ignoring them may be catastrophic.

This book owes its existence to a textbook entitled *Atmospheric Change: An Earth System Perspective*, which we prepared for science undergraduates in colleges and universities. We had intended, early in its development, to write the book so that it could serve a lay audience as well. This eventually proved to be impossible, and we have instead used the textbook as a resource from which to draw in writing the present volume.

A number of our scientific colleagues read portions of early drafts of the present material. We are especially grateful to Russell Dickerson and Alex

Pszenny, who reviewed the entire manuscript. Many of the staff at W. H. Freeman and Company have provided important, often invaluable, service. We are especially pleased to single out Jonathan Cobb, under whose overall guidance this project took form, Moira Lerner, whose fierce determination to make each concept understandable and each sentence proper has added greatly to the clarity of the presentation, Travis Amos, who labored to find just the right visual images to illustrate our ideas, and Diane Cimino Maass, who guided all the pieces into final form as project editor.

As always, we thank our wives, Susannah and Terttu, for their love and support.

The atmospheric measurements station at Cape Grim, Tasmania.

Taking the Pulse of Earth

Civilization exists by geological consent, subject to change without notice.
—*Will Durant*

Cape Grim, at the western edge of Tasmania, was well named by the early mariners who attempted to avoid its shoals and rocks. Today, perched on a Cape Grim cliff, an atmospheric measurements station of the Commonwealth Scientific and Industrial Research Organisation of Australia receives air that has traveled over 5000 kilometers (km) of southern ocean. Since 1976, concentrations of atmospheric chemicals have been measured here at regular intervals.

Halfway around the world from Cape Grim is Mauna Loa, a volcanic peak on Hawaii's largest island. Upon its western slope sits an atmospheric measurements station operated jointly by the Scripps Institution of Oceanography in La Jolla, California, and the U.S. National Oceanic and Atmospheric Administration. On most days, the wind is from the west, bringing air that has crossed half the Pacific Ocean on its passage from China. Carbon dioxide and other gases have been monitored at Mauna Loa since 1958.

Cars, buses, and motorized cycles jockey for space on the crowded streets of a university suburb of Mainz, Germany. Above them, students of the Max Planck Institute for Chemistry assemble on a rooftop to examine the readings of instruments measuring sulfur and nitrogen oxides, ozone, and other urban air pollutants, as well as wind speed, wind direction, and rainfall.

Working in offices at the University of East Anglia in Norwich, England, researchers organize and analyze temperature measurements from hundreds of locations around the world, deriving average hemispheric and global temperatures to determine whether the planet is cooling or warming.

At Cape Grim, evidence accumulates that the concentration of small atmospheric particles is increasing. At Mauna Loa, atmospheric carbon dioxide concentrations show not only an annual cycle corresponding to plant photosynthesis and respiration, but an underlying upward trend related to the global use of fossil fuels. At Mainz, urban ozone is showing an increase. In Norwich, analyses of historic temperature measurements show Earth to be about half a degree centigrade warmer than a century ago. As the pace of investigation has intensified over the past few decades, scientists have become increasingly aware that such phenomena are not independent attributes of continent, ocean, and sky, but the vital signs of an interconnected system. We are all familiar with how physicians approach the human system: they measure pulse rate, cholesterol level, and motor function as clues to the state of the system and to the rates of change of its properties. In just such a way, today's geoscientists monitor what the British scientist James Lovelock has termed "the physiology of Planet Earth."

The Earth System Perspective

Systems of all types share a few common characteristics, the most basic of which is that each is a group of interacting, interrelated, or interdependent elements forming a collective entity. Another characteristic of systems is that they are dynamic, reacting constantly to driving forces and perturbations from within and from without. A consequence of this dynamism is that the longevity of a system is not assured: some systems remain virtually un-

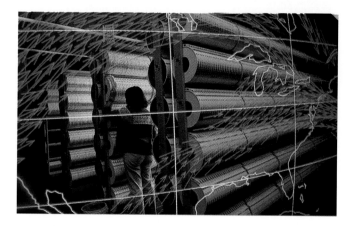

A climate forecast for the North American continent, as produced by the meteorological center at Reading, England. Massive amounts of atmospheric information are needed to produce the computer calculations that generate the forecast.

changed for long periods, others alternate rapidly between growth and decline, some sicken and die. Our view of Earth as a dynamic, interdependent system is supported in part by solid evidence that the planet has seen all these stages except the last, and perhaps even that stage can be predicted.

Developing scientific approaches to imperfectly understood systems has been the central focus of ecosystem ecologists, who attempt to understand ecosystems (the living organisms that make up an ecologic community together with the physical environments that they occupy) by evaluating how each element functions as a part of the greater whole. They also attempt to understand how entire ecosystems evolve and age, as perhaps happens in all cases but certainly occurs if external conditions change.

If Earth, then, is properly labeled a system, it should be possible to describe its processes in the same terms that ecosystem scientists use, and indeed we can do so. Its energy sources (that is, its driving forces) are solar radiation and, to a substantially smaller degree, the heat lost gradually from Earth's hot inner core. Earth's primary digestive system, if we may so term it, is its plant life, which transforms a fraction of the energy that is supplied into useful forms such as leaves and seeds. And, just as ecosystem elements such as birds or elephants adjust their immediate physical states by eating food or seeking shelter, elements of the planetary system such as oceans, deserts, and tectonic plates adjust their physical states by such processes as the evaporation and condensation of water and the exchange of minerals.

Ecosystems often respond quickly and dramatically to changes in energy sources, as the Earth system does to fluctuations in solar radiation. Hence, we are here concerned with possible changes in solar radiation, with the responses of the receptors of that radiation, such as ice caps and maple trees, and with the factors that influence those receptors. Ideally, we hope to study and predict those changes on all time and space scales that have a bearing on the system's behavior. In practice, of course, the facets of the Earth system that are easiest for us to study are those that operate on the time and space scales most like our own. The most obvious such system element is climate.

What exactly is meant by the term "climate"? The usual definition is that climate is the average condition of the weather over several decades, as exemplified by such characteristics as temperature, wind velocity, and precipitation. Climate is influenced by many factors: the heights of mountain ranges, the slowly changing locations of the continents and oceans, the ocean currents, clouds in the atmosphere, the extent of polar ice caps, the density of vegetation, and so on.

Temperature, air motions, and climate are influenced by five different Earth system regimes with widely varying impacts and time scales. The atmosphere is one of the five; the others are the biosphere, the hydrosphere, the cryosphere, and the pedosphere. The atmosphere, by far the most rapidly varying, responds quickly to external forces, such as daytime heating and nighttime cooling. It is the regime to which we as human beings are most directly exposed. A key property of the atmosphere, one that determines many others, is its pressure, which is highest at Earth's surface and decreases rapidly with increasing altitude by about a factor of 2 for each 5 kilometers (km). Another crucial atmospheric property is the temperature; unlike the pressure variation, the temperature variation with height is quite complex. Atmospheric scientists use the altitudes at which temperature changes abruptly to distinguish different regions for study and reference, as

shown in the diagram below. Beginning at Earth's surface, these regions are called the troposphere, the stratosphere, the mesosphere, and the thermosphere, and their boundaries the tropopause, the stratopause, and the mesopause, respectively. In this book, we will restrict our discussions to the troposphere and stratosphere, which are the most important regions for climate and life on Earth. Both regions are strongly affected by anthropogenic, or man-made, and natural emissions at the surface. In addition, the stratosphere is affected by volcanic explosions, aircraft emissions, and solar eruptions.

Closely linked to the atmosphere is the biosphere, the aggregate of plant and animal life on Earth. Seasonal changes in vegetation affect the albedo (the degree of absorption of solar radiation) of a geographical region, as well as its water budget. As part of the biosphere ourselves, we have wrought changes on it such as deforestation, agri-

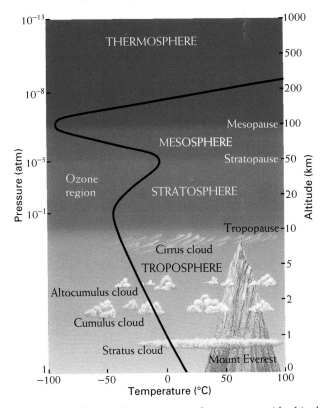

The variation of atmospheric pressure and temperature with altitude above Earth's surface. The regions of the atmosphere are noted, and the Himalayas are drawn in for perspective.

culture, and urbanization that can also have profound effects on climate locally, regionally, and globally.

The hydrosphere, comprising all liquid waters of Earth, influences temperature and circulation on time scales of seasons to centuries. The oceans are crucial to climate because they absorb the bulk of the solar radiation that falls upon Earth; the absorbed energy vaporizes water, which ascends into the atmosphere and later condenses into clouds, releasing the absorbed energy as heat. Ocean currents transfer that heat from the tropical regions, where the sun is most intense, to the polar regions.

The cryosphere is not defined by its material content but by a physical characteristic: it is the portion of Earth's surface with average temperatures below the freezing point of water. The bulk of the cryosphere is located at or near the poles, but on several continents cryospheric regions are also found atop high mountain ranges. Snow and ice are much better reflectors of solar radiation than uncovered land and sea, and they cause a substantial decrease in surface heating. Cryospheric changes occur seasonally, but major variations in the cryosphere have time scales ranging from centuries to millennia.

The slowest-acting region is the pedosphere, the solid portion of Earth's surface. The pedosphere rides on continental structures that evolve over time periods of millions of years as a consequence of tectonic plate motions. Continents covered with glaciers reflect much more solar radiation than the oceans do, so those past geological periods in which the continents were located primarily at high latitudes rather than near the equator have been periods during which the planet's climate tended to be much cooler than average.

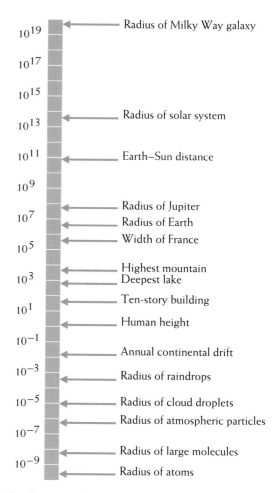

Spatial scales in Earth system science. The ordinate is logarithmic, so each step up from the bottom represents a distance (in meters) 10 times that of the segment just below it.

Scales of Space and Time

The components of the Earth system include atoms and continents, and their rates of change encompass eons and femtoseconds. Both spatial and temporal scales in Earth system science thus present an enormous range—in the extreme a difference of more than 20 orders of magnitude. It may not be possible for human beings whose own experiences span very limited fractions of these ranges to intuitively relate to the extremes, but even the attempt to do so will be rewarded with improved perspective.

First, consider information relating to size. The smallest spatial scales, of the order of a tenth of a billionth of a meter, are those of atoms and molecules; this is the sort of distance on which chemical reactions occur and which applies to the basic processes that change the atmosphere and climate. Larger than atoms and molecules, but still measuring small fractions of a millimeter, are the components of the atmosphere's solid and liquid water phases—particles, cloud droplets, and raindrops. The typical annual movement involved in continental drift is a hundred times greater, a centimeter or two per year; thus the procession of the continents has about the same dimension as that of a large snowflake. Lengths expressed in meters (m) are typical of human beings and their structures: a basketball player stands 2 m high, the minarets of the Taj Mahal rise 40 m above the ground. Much larger lengths, several hundreds to thousands of meters, describe major physiographic features—deep lakes, ocean depths, and tall mountains (Mt. Everest is 8872 m above sea level). The dimensions of countries, continents, and planet Earth itself are of the order of 100 to 10,000 km. The distance from Earth to the Sun is 150 million km and the radius of the solar system is a thousand times larger still. Beyond that extend the far reaches of our own galaxy, the Milky Way, and the universe itself. These realms might seem remote from our experience of atmosphere and climate. But, traveling distances so great they must be measured in light-years, galactic cosmic rays are responsible for the production of both natural radioactivity and nitric oxide in Earth's upper atmosphere.

Ordering these spatial domains by the widely varying scales that measure them, as shown in the figure on the facing page, reveals a key relationship: the scale of the physiographic features of Earth is very large, whereas the scale of the processes that drive the Earth system is very small. The creation of drops of acid from nonacidic reactants, the weathering of rocks, the eruption of volcanoes, and the formation and dispersal of clouds are all products of vast numbers of individual physical and chemical processes, each acting on a tiny scale. We cannot hope to understand the large, obvious features and dynamics of Earth unless we appreciate the minute but vital processes that bring them about.

Although many of these spatial scales are outside our common experience, it is nonetheless possible for us to observe them, using an electron microscope for the very small sizes, for example, and a telescope for the very large. More daunting conceptually are time scales beyond our direct experience. The fact that these extreme time scales are veiled from us by blinding speed or snail-like lethargy obscures our perception of many important changes in atmosphere and climate. For example, intuition encourages

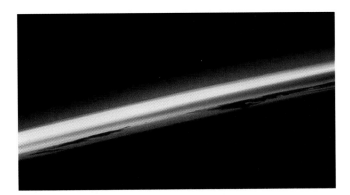

Compared with the diameter of Earth, the atmosphere, seen here in a photograph taken from the Space Shuttle Atlantis *in August 1992, appears thin and fragile. The blue light is scattered solar radiation, while the grayish-red band is a residue of the Mt. Pinatubo volcanic eruption of June 1991.*

us to think of the chemical composition of the air and the seas as being among those few constant properties of a changing world. It is now clear, however, that atmosphere and oceans must be regarded, like Earth itself, as having evolved through a number of different stages over a long period of time. This evolution is the result of variations in chemical cycling among the atmosphere, the solid surface, and the oceans, of the movement of continents and of continental modification caused by volcanic eruptions, of changes in the intensity of solar radiation, and of the interplay of the atmosphere with flora and fauna. Therefore, it is necessary to understand how these factors have evolved since the beginning of the planet. Furthermore, intermittent impacts of meteorites are thought to have led to major perturbations in the chemical composition of the atmosphere and, as a result, in Earth's climate, with major consequences for the evolution of life on Earth.

Is it possible to know how the properties of the atmosphere—which German astronaut Ulf Merbold described as "a fragile seam of dark blue light"—have changed over the eons of existence and through the agency of gases, particles, and droplets? Can we tell what the consequences of those changes might be for the Earth system? As we might expect, it is possible to answer these questions with some assurance for the current century and with progressively less certainty for earlier times. In the opposite direction of time's arrow, we can be reasonably sure of the answer for the immediate future and much less certain for the far future.

Scientists trace the existence of all matter and energy back to the Big Bang, a rapid expansion of primordial matter into space some 15 to 20 billion years ago. As the matter cooled, random motions coalesced the denser portions into stars, galaxies, and planetary systems. At least, astronomers *suspect* the existence of many planetary systems throughout the Milky Way and the universe, but we know for certain of only one—our own.

Radioactive dating of isotopes in meteorites and moon rocks, together with observations of star formation elsewhere in the Milky Way, make it apparent that the Sun, Earth, and the rest of the solar system were created by the gravitational collapse of a huge nebula of dust and gas between 4.5 and 4.7 thousand million years ago (that is, 4.5 to 4.7 gigayears before the present, abbreviated as Gyr BP). The oldest rock grains on Earth have been radioactively dated at about 4.2 Gyr BP, and the dates of different rock sections and the sediments in which they were identified serve to define for geologists the epochs in Earth history. Some of the events that distinguish the geological divi-

The stability of climate over several millennia has permitted land to be cleared, developed, and cultivated so as to produce the abundance of food needed by a growing population, such as in these fields surrounding a village in the Yorkshire Dales of England.

sions in time appear in the figure on the facing page. The time scale is not linear because the more abundant evidence of Earth's recent history has allowed scientists to develop a much more detailed description for the recent past than is possible for the more distant past with its sparser data base.

A few major geological and paleontological events are crucial to the study of the relationships among Earth's solid surface, its oceans, its atmosphere, and its plant and animal life. Perhaps the most significant points in the time scale are the formation of Earth; the appearance of the earliest known life, at about 3.8 Gyr BP; the Proterozoic–Cambrian transition, when the increasing concentrations of atmospheric oxygen permitted a great explosion of diverse life forms; the Permian–Triassic transition, when the continents as we know them first took form; the Cretaceous–Tertiary transition, when the great extinctions of dinosaurs and other life forms occurred; and the Holocene era, the 10,000 years that encompass the recorded history of humanity. We will refer again and again to events on the geological time scale.

Stability, Metastability, and Instability

Why *did* the dinosaurs die out? Did climate undergo a drastic change? Why is Earth's average temperature warm enough to keep water liquid? Could that situation change, and how rapidly might such a change occur? Evidence that major changes have occurred in the past suggests that the Earth system may not be indefinitely stable. Rather, it may oscillate among several different states of climate, with attendant impacts on dinosaurs, plants, marine microorganisms, and all other life forms—including human beings.

Although assessing the stability of complicated systems can be a demanding task, the concepts of stability and instability are fairly straightforward, especially when described by analogy with a mechanical system. Consider such a system at equilibrium. The sum of its forces is zero and is characterized by the gravitational potential of a resting object. Now consider what happens to the system if a small perturbing force is applied. A ball placed in a deep depression (position 3 in the figure on page 8) and

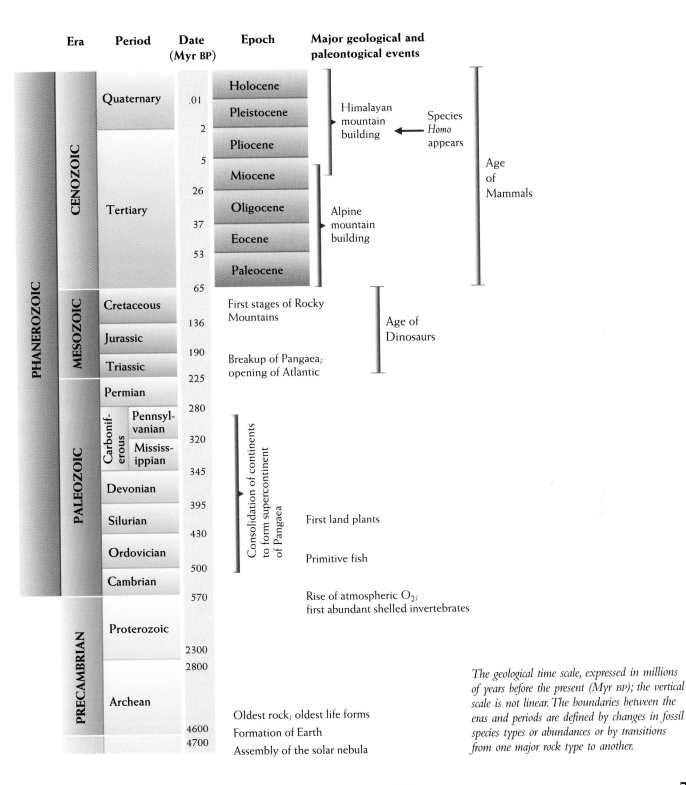

Era	Period	Date (Myr BP)	Epoch	Major geological and paleontogical events

PHANEROZOIC

CENOZOIC
- Quaternary
  - .01 — Holocene
  - Pleistocene
  - 2
  - Pliocene
  - 5
- Tertiary
  - Miocene
  - 26
  - Oligocene
  - 37
  - Eocene
  - 53
  - Paleocene
  - 65

Himalayan mountain building

Species *Homo* appears

Alpine mountain building

Age of Mammals

MESOZOIC
- Cretaceous
  - 136
- Jurassic
  - 190
- Triassic
  - 225

First stages of Rocky Mountains

Breakup of Pangaea; opening of Atlantic

Age of Dinosaurs

PALEOZOIC
- Permian
  - 280
- Carboniferous
  - Pennsylvanian
  - 320
  - Mississippian
  - 345
- Devonian
  - 395
- Silurian
  - 430
- Ordovician
  - 500
- Cambrian
  - 570

Consolidation of continents to form supercontinent of Pangaea

First land plants

Primitive fish

Rise of atmospheric O_2; first abundant shelled invertebrates

PRECAMBRIAN
- Proterozoic
  - 2300
  - 2800
- Archean
  - 4600
  - 4700

Oldest rock; oldest life forms
Formation of Earth
Assembly of the solar nebula

The geological time scale, expressed in millions of years before the present (Myr BP); the vertical scale is not linear. The boundaries between the eras and periods are defined by changes in fossil species types or abundances or by transitions from one major rock type to another.

Taking the Pulse of Earth

then pushed will soon stop moving; it is *stable*. But if started rolling down a hillside of steady slope (position 2), the same ball will not stop rolling; it is *unstable*. A small force applied either to the right or left of a ball in position 2 is sufficient to move it to another state. If a ball is placed in position 1 and a small force is applied to it, the ball will oscillate for a while and then settle back into its original position; however, if a larger force is imposed, the ball may be moved into an unstable state. Position 1 is thus described as a *metastable* state. A ball in position 3 may still be metastable, but is less likely to be displaced than if it were in position 1 or 2; it will remain where it is in the absence of the application of a large perturbing force. Should position 3 be the lowest possible energy state, it is called the *stable state*.

The concepts of stability and instability can also be looked at in a more general way. In the figure on the facing page, we illustrate systems with four types of stability characteristics, again using gravitational potential as a paradigm for stability. The instability increases from bottom to top. The concept of geometric figures on hilly terrain allows us to picture various combinations of states and sensitivities. Sensitivity is a measure of the intrinsic resistance of a system to change; we indicate this property

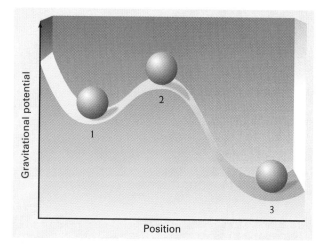

Stability diagram for a mechanical system. The curved surface indicates the boundary of the "states" allowed by the system. The ordinate parameter is the gravitational potential energy. Position 1 is a metastable state, position 2 an unstable state, and position 3 a stable state.

schematically here: a hexagon is more stable on a hillside than is a ball, and a cube is more stable still. Whether or not a system is stable to a modest perturbation is seen to be characteristic not only of the allowed energy paths of the system, but also of the object's position on the hillside.

How do stability and instability enter into discussions of Earth system science? The answer to that question depends on the particular system under consideration, on the strength and type of perturbation, or "forcing" (which might be, for example, increases in solar radiation intensity or changes in atmospheric composition), and on the system's sensitivity (for example, an equal addition of heat causes a greater temperature change in the atmosphere than in the ocean). The time scale is also important: the bed of a river may be quite stable over a few decades or perhaps a few centuries, but it assumes a new state if it is allowed to evolve for a few millennia. For example, the main channel of the River Rhine in the Netherlands was about 30 km north of its present location during the Roman times; the small river that currently occupies that channel is called the Oude Rijn (Old Rhine). On longer time scales, such as during the Pleistocene and Holocene epochs, global climate, as measured by the area covered by the polar ice packs, swung from glacial to interglacial states a number of times in each million years. Climate, therefore, appears to be of stability type C, a system with several different metastable states. Some natural climatic entities, such as the flow patterns of upper atmospheric winds, are so easily perturbed that they must be regarded as very sensitive. In a sense, this book is a story of Earth's possible positions within various stability diagrams. However, the Earth system is much more complex than mechanical systems, and we are often uncertain of properties equivalent to different types, sensitivities, and topologies. Moreover, as the Earth system evolves, its stability states may be changing. For example, continental drift and mountain building may influence the stability of Earth's climate. The perturbing forces may either be internal to Earth's system (volcanic eruptions, tectonic activity, human industrial and agricultural activity) or external (solar radiation variations, impacts by meteorites, supernova explosions). A dramatic example of change wrought by an internal instability is the appearance of oxygen as a major constituent of the atmosphere some 1.5 Gyr BP, owing to

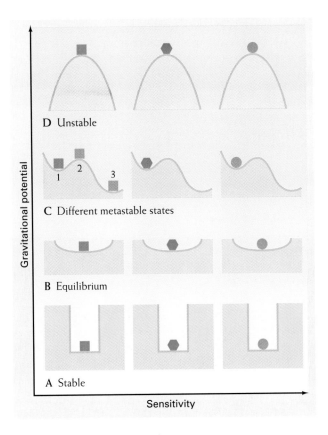

Gravitational potential

Sensitivity

D Unstable

C Different metastable states

1 2 3

B Equilibrium

A Stable

Conceptual diagrams of stabilities for systems of different types and sensitivities. Instability increases from bottom to top; sensitivity to perturbation (suggested by the shapes of the figure) increases from left to right (at least for B, C, and D). The meanings of "small" and "large" are functions of the sensitivity of the system.

A. Stable. Nothing except destruction of the system can cause a change of state.
B. Equilibrium. A perturbation can temporarily modify a state, but the system can never move far from its position on the vertical axis.
C. Different metastable states. If a large enough perturbation is applied, then a new metastable state is achieved.
D. Unstable. A small perturbation causes a radical change of state.

the introduction of photosynthesis by biological evolutionary processes. This "natural pollution" led to the demise of much of the then-existing biosphere, which was based on anoxic (oxygen-poor) processes.

Regardless of the stability type of today's Earth system (the preliminary evidence would suggest type C), we certainly cannot specify the system's current position within that regime. A stability transition could occur, for example, because of human activities affecting climate and chemistry and, hence, global-scale processes such as the atmospheric energy budget. Studying Earth's physiology in its various forms helps us understand the state of today's system and guides us in formulating computer models that help us probe the possibilities for tomorrow's. The crucial

realization, however, and one that will take the remainder of the book to develop fully, is that if the Earth system has never before passed from a stability of position 1 or 2 to one of position 3, we have no way of knowing what the characteristics of the state represented by position 3 may be, or what kind of forcing may encourage the Earth system to enter that state or leave it. This incomplete understanding in the face of nature's awesome capacity for change argues for a cautious approach to human perturbations of the Earth system. Hence, even though our knowledge is incomplete, we should not be reluctant to make decisions about the prudent use of the planet's resources. We can no longer afford to defer such important decisions to future times and coming generations.

*Weather and climate are not only a continual influence on all forms of life on Earth, but are the
most obvious manifestations of all the processes of the Earth system.*

Agents of Climate 2

Nothing that is can pause or stay;
The moon will wax, the moon will wane,
The mist and cloud will turn to rain,
The rain to mist and cloud again,
Tomorrow be today.
—Henry Wadsworth Longfellow

"**D**id you remember your umbrella?" This question, commonly asked by young and old, rich and poor, healthy and infirm, is all the evidence we need to prove that weather and its variability are considerations in everyone's life. Weather obviously varies day to day, and even hour to hour, but these variations are superimposed upon more general weather patterns occurring over long time periods. Before we can discuss the weather's variability and stability, however, we need to describe the current condition: the average over several decades of common atmospheric factors such as temperature, humidity, precipitation, and the like—a package of information that we call *climate*.

Why is it that year after year, we expect fog in London in February, sunshine in Addis Ababa in June, snow in Minneapolis in November? These climatic behaviors reflect the conditions and motions of the atmosphere, and most people know some of the key principles that govern them: temperature changes with latitude (equator to pole) and with height (Mojave to Everest), surface winds blow in more or less regular patterns (the trade winds used by mariners), winds aloft blow in standard patterns as well (the jet streams, discovered and utilized by pi-

11

lots in World War II), and clouds form and rain and snow fall preferentially in certain locations and in certain seasons. One might expect an underlying driving force that explains all these patterns, and there is one: the radiation from the Sun.

The Radiant Engine

Except for the relatively small amounts of energy provided to Earth's surface from internal radioactive decay—energy that nevertheless moves continents and creates volcanoes, yet is less than a thousandth of the magnitude of the solar flux at the surface—the stimulus that drives the processes we call climate is the electromagnetic radiation energy from the Sun. Scientists conceptualize electromagnetic energy as a stream of photons (packets of energy) that may behave either as waves or as particles depending on the way they are being observed. Waves of electromagnetic energy are characterized by their wavelength (λ), the physical distance between successive crests, and frequency (ν), the number of crests that pass a given point per second. According to a theory proposed by the German physicist Max Planck in 1900, the energy of a photon is calculated as

$$E = h\nu = h/\lambda \qquad (2.1)$$

where the quantity h, known as Planck's constant, is a universal constant of proportionality.

Heated masses of material, such as the Sun or one of the planets, emit ensembles of photons over a broad band of wavelengths, a band whose location in the electromagnetic spectrum depends on the mass's temperature. The hotter the radiating body is, the more photons it emits and the more the maximum number of emitted photons shifts in energy toward the more energetic, shorter wavelengths. The Sun, with an effective surface temperature of more than 5000°C, far hotter than the melting points of gold and iron, emits most of its energy at a wavelength band centered at about 500 nanometers (1 nm = 10^{-9} m), which is yellow-orange in the visible spectrum. Some radiation also occurs in the invisible high-energy ultraviolet range, and some in the lower energy, invisible infrared domain.

This solar energy, unimpeded and untransformed, streams through the vacuum of space at 300,000 kilometers per second (km s^{-1}). Shakespeare described air as "this majestical roof, fretted with golden fire." That golden fire is a rain of solar photons, which upon arrival at Earth encounter an atmosphere that is not a vacuum but an envelope of gaseous matter. The interactions of solar radiation with the atmosphere's gases and with Earth's surface begin the energy transfers that ultimately bring about climate.

Radiative energy transfer involving individual atoms and molecules in the atmosphere, as opposed to a compact, heated mass of material, is accomplished by many individual transitions between two of the energy states permitted by atomic or molecular structure. For short-wavelength solar radiation, the most important atmospheric absorbers are molecular, or diatomic, oxygen (that is, O_2) and ozone (O_3). The former absorbs photons with wavelengths shorter than about 240 nm, well into the invisible ultraviolet range. This absorption process uses the photon's energy to break apart the bonds holding the oxygen atoms together:

$$O_2 + h\nu \rightarrow O + O \qquad (2.2)$$

(In such equations $h\nu$ is the standard symbol for a single photon of appropriate frequency to be absorbed by, and furnish energy to, the molecule destined for dissociation.) The free oxygen atoms (O) that are produced react with other oxygen molecules to form ozone,

$$O + O_2 + M \rightarrow O_3 + M \qquad (2.3)$$

where M denotes any air molecule, usually molecular nitrogen (N_2) or oxygen, that acquires the excess energy liberated by reaction 2.3 and disperses it to surrounding molecules by colliding with them. In this way, the newly formed ozone molecule is denied the excess energy that would stimulate it to revert to O and O_2. Ozone preferentially absorbs photons of somewhat longer wavelengths than O_2 does, but with similar results:

$$O_3 + h\nu \rightarrow O_2 + O \qquad (2.4)$$

The excess energy contained in the reaction products is again dispersed into the atmosphere by collisions among molecules. Thus, each absorption of a photon by ozone provides a small amount of heat to that portion of the atmosphere near where the absorption occurred.

This small sampling of atmospheric photochemistry (that is, chemistry involving photons) can be used to describe how solar radiation organizes the structure of the atmosphere. Most solar radiation with wavelengths shorter than about 100 nm is absorbed by N_2, O_2, N, and O before arriving within 100 km of Earth's surface. Photons with wavelengths longer than 100 nm can penetrate more deeply into the atmosphere. Here, the strong absorption by O_2 limits photons of wavelengths shorter than about 200 nm to an altitude of 50 km and higher. Photons with wavelengths longer than about 200 nm penetrate further, to 25 to 30 km on average, and for them O_3 assumes the role of major absorber. This ozone absorption provides the energy that heats the stratosphere and much of the mesosphere. It also screens Earth's surface from the photons just short of 310 nm in wavelengths that are responsible for biological mutations, sunburn, and the Sun's other harmful physiological effects. Hence, a decrease in ozone concentrations in the stratosphere leads to an increase in the intensity of the most energetic radiation reaching Earth's surface. Were stratospheric ozone to be collected at sea level temperature and pressure, it would be only about 3 mm thick—not much material to do such an important job. In contrast to these shorter-wavelength photons, screened so effectively by Earth's atmosphere, photons with wavelengths longer than about 310 nm are only modestly attenuated by air molecules, clouds, and particles. It is this radiation that illuminates the planet.

The portion of solar radiation that is not absorbed during its passage through the atmosphere—about half of the radiation that began the atmospheric journey—heats

The National Center for Atmospheric Research in Boulder, Colorado. In this beautiful building, designed by I. M. Pei and built from red sandstone to blend with the surrounding mountains, atmospheric scientists study the processes responsible for weather and climate.

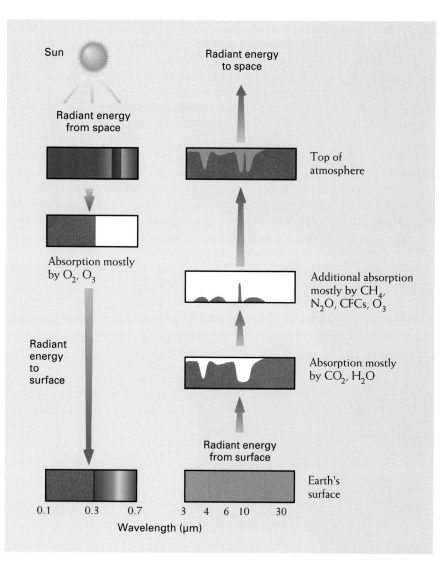

Sun

Radiant energy
from space

Absorption mostly
by O_2, O_3

Radiant
energy
to
surface

Radiant energy
to space

Top of
atmosphere

Additional absorption
mostly by CH_4,
N_2O, CFCs, O_3

Absorption mostly
by CO_2, H_2O

Radiant energy
from surface

Earth's
surface

0.1   0.3   0.7   3 4 6 10   30

Wavelength (μm)

The incoming radiation from the Sun (upper left) is of different wavelengths and is partially absorbed by atmospheric molecules. The portion not absorbed by the atmosphere is predominantly visible radiation (lower left) that warms Earth's surface, and the warmed Earth emits heat radiation (lower right). Some of the latter is absorbed in the troposphere by water and carbon dioxide, thus retaining the heat there. Other gases, mostly emitted from human activities, absorb some of the transmitted radiation in the lower stratosphere. The remainder escapes to space. All absorbing gases are termed greenhouse gases and render Earth some 33°C warmer than it would otherwise be without them.

the soil, water, and vegetation upon which it falls. Because any body warmer than a temperature of absolute zero, where all molecular motion ceases, will radiate, Earth emits radiation to the atmosphere. Earth is much cooler than the Sun, so most of its radiation occurs at long wavelengths, in the invisible infrared portion of the spectrum. Like the incoming solar radiation, a portion of

the outgoing radiation is absorbed by atmospheric molecules. The energy of the infrared photons is insufficient to cause chemical changes. Instead, the absorption of these photons merely increases the internal vibrational and rotational energy of the absorbing molecules. That excess energy is subsequently transferred to the atmosphere as kinetic energy (heat) by molecular collisions, and so Earth, warmed by the Sun, in turn warms the atmosphere.

The transfer and transformation of radiation from above and below the atmosphere is outlined in the figure on the facing page. The *incoming* radiation is effectively absorbed by small atoms and molecules if its wavelength is shorter than about 0.3 micrometers (300 nm) and transmitted if its wavelength is longer. The *outgoing* radiation, of much longer wavelength, is efficiently absorbed only by larger, less symmetric molecules. By far the most important of the latter are water vapor and carbon dioxide. They are sufficiently abundant that they strongly reduce the transmission of radiation to space in many wavelength regions, particularly in much of the 3 to 30 micrometer region (1 μm = 10^{-6} m), where the principal radiation from Earth occurs. Terrestrial radiation has the best chance of escaping to space if it has wavelengths where H_2O and CO_2 absorb inefficiently, in the so-called infrared windows. Several such windows are shown in the figure, the most crucial of which is from 8 to 12 μm, because this spectral region includes most of Earth's emitted photons. The escape of many of these photons into space prevents their energy from heating the planet. However, any atmospheric trace gas that happens to absorb efficiently in the 8 to 12 μm window is able to retain those photons and add their energy to the atmosphere as heat. Several of the gases emitted prolifically by human activities happen to absorb in just this wavelength region: methane (CH_4, emitted from agricultural activities, natural gas leaks, and landfills), nitrous oxide (N_2O, emitted from agricultural activities and industrial processes), and chlorofluorocarbons (CFCs, emitted from air conditioners, foamed insulation, and pressurized containers). Collectively, the absorbing molecules are termed greenhouse gases.

The atmospheric temperature structure pictured in the first chapter (page 102) can ultimately be explained by the emission and absorption of radiation. The troposphere, for example, is heated from below. Within this region, convective processes cause the heated air at the surface to rise into less dense air, which in turn causes the heated air to expand and cool. Thus the temperature decreases with altitude in the troposphere, to about −50°C at altitudes around 10 km at middle to high latitudes and to about −80°C at 17 km in the tropics. The increase in temperature in the stratosphere begins at the tropopause, caused by the increasing importance of stratospheric ozone's absorption of downwelling ultraviolet radiation from the Sun and by the absorption of upwelling infrared radiation from Earth. Because the product of the intensity of the solar ultraviolet radiation and the abundance of ozone relative to all air molecules (the mixing ratio) reaches a maximum at an altitude of about 50 km—that is, at the stratopause—maximum temperatures develop at this altitude. Above that altitude the mixing ratio of ozone decreases, causing relatively less energy uptake and heating. As a result, temperatures decrease in the mesosphere as altitude increases. In the thermosphere, the region farthest from Earth's surface, temperatures again increase rapidly with altitude, owing to molecular oxygen's absorption of solar ultraviolet radiation; at still higher altitudes, molecular nitrogen, atomic oxygen, and atomic nitrogen become the principal photon absorbers. The result is an atmospheric structure that is anything but simple; its changes in temperature with altitude form the backdrop upon which atmospheric motions occur.

Earth System Budgets

A detailed summary of how the radiation processes influence climate can be constructed in the form of a radiant energy "budget." This resembles a financial budget and balance sheet on which one either estimates or records actual values of the income and expense for a selected period. If income and expense are equal, then the budget is in balance. If income exceeds expense, then the monetary reserves increase. If expense exceeds income, then the monetary reserves decrease. Scientists use an analogous approach to keep track of such resources as energy, water, and molecules in the Earth system. An important difference, however, is that in an Earth system budget all of the input and output processes important for the system may not have been identified—or if they have, their magnitudes may be uncertain. Imagine a tub receiving water from several faucets and losing water through a number of drains of different sizes. When the water is supplied at constant (but probably different) rates by all the faucets and is removed at an equal total rate by drains (each probably having different flows), the water level remains constant. When the tank is very large, however, and the water exhibits some wave motion, it can be difficult to tell whether its absolute level is changing and whether the system is in balance or not. In that case, an observer might try instead to measure the rate of supply from each of the faucets and the rate of removal by each of the drains over a period of time to see whether their sums are equivalent. Alternatively, one can determine the pool size (the quantity of water in the tank) and either the rate of supply or the rate of removal. The determination of changes in the pool size then gives information about rates that are difficult to measure. The process of estimating or measuring the supply and removal rates and checking the overall balance by measuring the amount present in the reservoir constitutes an Earth system budget analysis.

Suppose that the input from one of the sources is increased; that is, in our analogy, the flow from one of the faucets increases. Will the water level rise and keep rising? The answer depends on whether one of the drains can accommodate the additional supply, such as the trough drain at the right side of the tank in the figure. If no such drain is present, then the water level will indeed start to rise and will continue doing so until some new equilibrium level is reached. Conversely, if the outflow through a drain is enhanced for some reason—say, the removal of an obstruction—then the water level will lower in the absence of a corresponding increase in the supply.

All of the hypothetical circumstances mentioned above occur in actual Earth system budgets, and all such budgets involve the same concepts. One key concept is that of the reservoir, which is any entity defined by characteristic physical, chemical, or biological properties that are relatively uniformly distributed. Examples include the atmosphere as a whole, spatial sections of the atmosphere, or chemically defined subsets, such as the oxygen, carbon dioxide, or water vapor pools. A second important concept is that of flux, the amount of a specific material moving from one reservoir to another within a specific period of time. Examples include the rate of flow of solar energy into the upper atmosphere, the rate of evaporation of water from Earth's surface, the conversion rate of methane to carbon monoxide in the atmosphere, and the rate of transfer of ozone from the stratosphere to the troposphere. Third, we have sources and sinks, which are rates of creation or destruction, respectively, of a specific material within a reservoir. The photochemical production and destruction of ozone are examples. A closed system of connected reservoirs that transfers and conserves a specific material is termed a cycle.

Ideally, then, all Earth system budgets have the same three basic requirements as the budget for water in our imaginary tank: determination of the present level in the system (that is, the concentration of a single property such as energy or water content or chemical species), the sources, and the sinks. In principle, knowing any two of these three components permits determination of the third. The great difficulty is that any species of importance to the environment is likely to have several sources and sinks, and it is generally necessary to study each source and sink individually.

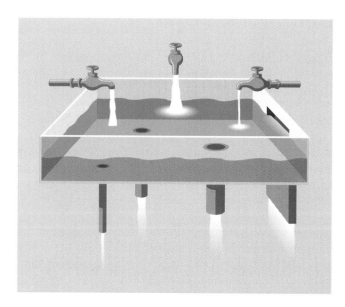

A simple conceptual system for budget calculations shows how the water level in the tub is determined by the water flows both in and out.

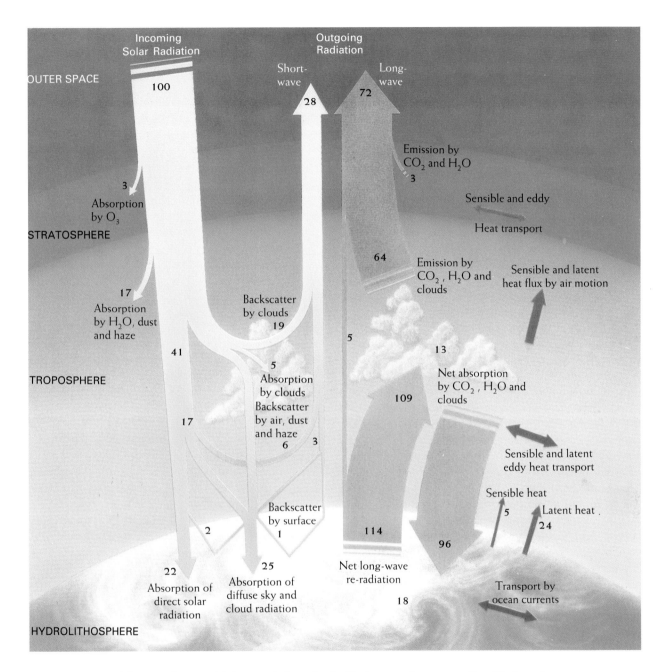

The annual mean global energy balance for the Earth–atmosphere system. Sensible heat is that heat transferred to the atmosphere from the heated surface by turbulent eddies; latent heat is that heat supplied to the atmosphere upon condensation of water vapor. The numbers are percentages of the energy from the incoming solar radiation.

The Energy Budget

The interplay of all these processes for radiant energy, the energy budget, is illustrated on the facing page, based largely on observations from Earth satellites of incoming and outgoing radiation over different regions of the planet. Of the total incoming solar radiation, an average of slightly less than 30% is returned into space, by reflection (backscattering) from clouds (about 19%), by backscattering by air molecules and particles in the air (together about 6%), and by reflection at Earth's surface (about 3%). Another 25%, approximately, is absorbed within the atmosphere, mostly by ozone in the stratosphere (about 3%), and by clouds (5%) and water vapor (17%) in the troposphere. The remaining 47% of the incoming solar radiation warms Earth's surface.

Of the solar energy absorbed at the surface, a little more than half is transformed into latent heat, that is, heat absorbed by water as a consequence of the water's transformation from liquid to vapor at Earth's surface. Latent heat is released again into the atmosphere when water vapor condenses into clouds. Other significant amounts of surface heat energy are transferred back into the atmosphere by convection and turbulence (about 10%) and by the absorption of Earth's infrared radiation by greenhouse gases. Compared to the 47% of the initial solar energy absorbed at Earth's surface, only a net 18% is lost to space by radiation. Carbon dioxide, water vapor, and clouds efficiently retain the remainder. As a consequence, of the 114 units of energy that are emitted (a unit is 1% of the energy arriving from the Sun at the top of the atmosphere), some 96 units are returned to heat the planet. This highly efficient cycling of energy between the atmosphere and the planet's surface produces the greenhouse effect, in which Earth's surface is about 33°C warmer than would otherwise be the case.

Careful study of the Earth system will show that the radiation budget is in balance: the outgoing radiation is exactly 100% of the incoming radiation. The planet achieves this balance by adjusting its own temperature through the processes discussed above. As with your own household budget, if input changes, output will change as well, or if the portion allocated to one part of the budget changes, another part will change to compensate. The stability of climate depends on the stability of the Sun's radiant energy flux and also on the stability of the energy absorption and emission processes in the atmosphere.

The Atmospheric Circulation

Until now, we might as well have been discussing an immovable Earth with a uniform surface, because nothing in the radiation budget analysis has depended upon other characteristics of the planet that we all know: its heating by the Sun is not spatially uniform, it rotates, and its surface varies in altitude and texture. These characteristics are crucial for climate, because they induce motions in the atmosphere and oceans. Those motions, together with a transfer of heat from surface to atmosphere that varies with geographical location, combine to produce the atmospheric circulation.

The most obvious of the location-dependent differences on Earth are quickly sensed from a world map. About 70% of the planet's surface is covered by the oceans, a striking property in a solar system without another water-covered planet and perhaps without another planet covered by liquid of any kind (although liquid methane may cover the surface of Titan, Saturn's largest moon). The continents, which make up about 30% of Earth's surface area, are distributed quite unequally over Earth, more than 65% of the land lying in the northern hemisphere. The land in the midlatitudes is generally the most suitable for agriculture and has become the most intensively developed; land near the equator is now coming under similarly heavy developmental pressure. Forested regions constitute about a third of the land area, most of the more heavily forested regions lying in either tropical or temperate latitudes. Grassland accounts for another quarter of the land area, mountains and deserts somewhat less. It is interesting that in this time of rapid population expansion, urban and suburban geography constitutes about a tenth of the total land area.

Superimposed upon this uneven distribution of land and land use is the variation in solar energy with latitude and season. More energy is received from the Sun at tropical (0 to 30°) and subtropical (30 to 40°) latitudes, for example, than is given off by outgoing terrestrial radiation. This is the case not only for land areas but for the oceans as well. As a consequence, there is year-round en-

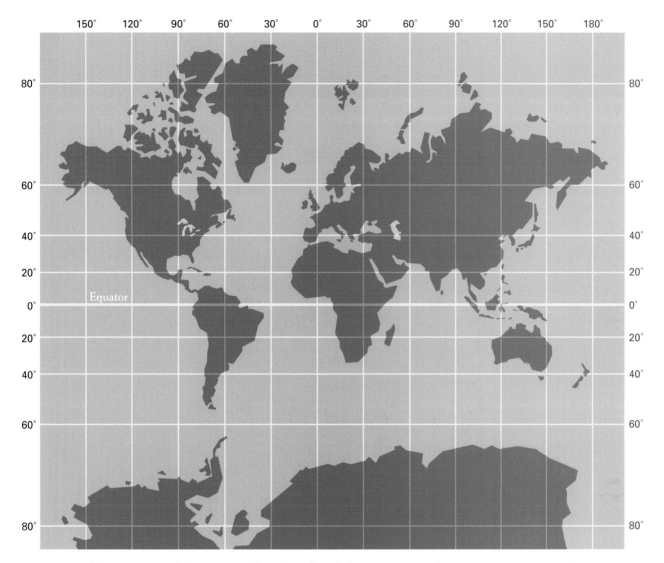

The positions of the continents and the oceans on the surface of Earth (Mercator projection).

ergy transfer from low to high latitudes by both air and ocean currents.

Because the heating of Earth is generally greatest over the equator, heated equatorial air expands and moves upward to a greater degree than does the air at other latitudes. As the equatorial air rises, air at low altitudes and from higher latitudes in both hemispheres moves toward the equator to take its place. This flow is balanced by a flow away from the equator at higher altitudes. While flowing poleward, the upper air cools by radiation to space. At about 30° latitude in both hemispheres, this cooled air is dense enough to descend, closing the Hadley circulation (named for the Englishman George Hadley, who first described Earth's circulation patterns in 1735) on both sides of the equator. The strong upward motion of air near the equator is characterized by heavy precipitation in

Illuminating a Planet with an
Inclined Axis of Rotation

*T*he amount of sunlight received by a specific point on Earth differs with latitude and season. These differences are reflected in variations of temperature with geographical location, and are crucial to an understanding of atmospheric motions, weather, and climate. The top diagram is the situation at the vernal equinox (March 21) and the autumnal equinox (September 21), when Earth's axis of rotation is perpendicular to the incoming solar radiation. Compare the radiation received by a unit area at point S in the top panel, the subsolar point, with that received at point P, a unit area at 66.5° north on the polar circle, and at point H, a unit area at the pole itself. Area S receives significantly more radiation than the unit area at point P, both because of the latitudinal variation and because at point P the scattering and absorption of a longer passage through the atmosphere reduces the flux more efficiently. At this time of year, all points on Earth except exactly at the poles have 12 hours each of daylight and darkness. In the northern hemisphere summer (center panel), the tilt of Earth increases the solar flux per unit surface area on the northern latitudes and increases the duration of day versus night; in the southern hemisphere summer, the opposite is true (bottom panel). The regions poleward of the polar circles have 24 hours of day at the summer solstice. Exactly at the poles, half of the year it is day (during spring and summer) and half of the year it is night (during autumn and winter). When solar radiation heats the surface, air near that surface is heated as well. Because of the many factors causing differential heating of Earth—different reflectivities of land and ocean, presence or absence of snow and ice, type of vegetative land cover, and so forth—regions of cooler air near warmer air are common. Winds and moving weather systems are nature's actions designed to smooth out these differences in the atmosphere's temperature and pressure.

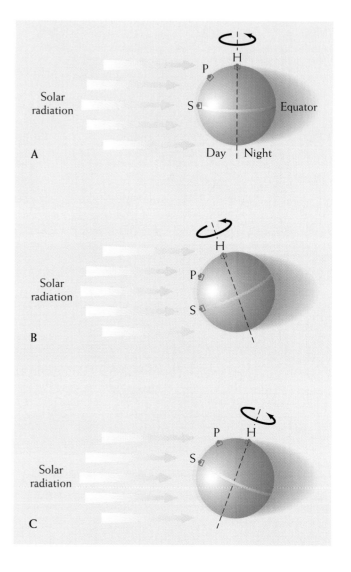

Solar radiation on a planet at different axial inclinations.

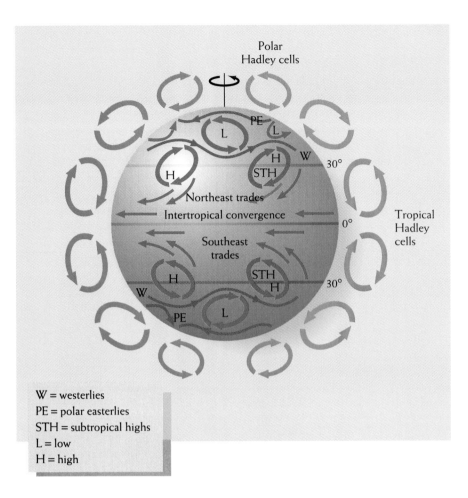

Polar
Hadley cells

PE

L

L

H W
STH 30°

H

Northeast trades

Intertropical convergence 0° | Tropical
Hadley
cells

Southeast
trades

STH
H

H 30°

W

PE L

W = westerlies
PE = polar easterlies
STH = subtropical highs
L = low
H = high

Principal features of the atmospheric circu-
lation. On the sphere, representing Earth,
the diagrams refer to surface winds. The
transient highs (H) and lows (L) we see
on daily weather maps are primarily lower
atmospheric features, extending only about
5 km into the atmosphere, whereas the
Hadley circulation can extend vertically as
much as 20 km from the Earth's surface,
through the troposphere and partially into
the stratosphere.

thunderstorms and by relatively low surface pressures. The latitudes where the air descends, on the other hand, are fair weather regions, with little precipitation and subsiding dry air that produces high surface pressures. These are the horse latitudes, so termed because merchant sailors occasionally threw overboard horses carried as cargo, in desperate attempts to lighten their load and use the available gentle breezes to reach their destinations.

A portion of the downward-moving air at the horse latitudes moves toward the equator to compensate for the up-

ward flow there. As a result of Earth's rotation, these airflows are deflected—to the right in the northern hemisphere and to the left in the southern hemisphere. The result in the north is the strong, reliable winds that were called trade winds by seventeenth century merchant sailors, who used them effectively to reach the Americas from Europe.

The general westerly flow of the atmosphere at temperate latitudes is often disrupted near the ground by the complicating effects of surface cooling and heating and by the presence of mountain ranges. Further aloft, the westerly flow is steadier in direction and speed, reaching a maximum velocity often exceeding 100 meters per second ($m\ s^{-1}$) in the jet streams at altitudes of 10 to 12 km. It is these streams that hasten the travel of many an airliner.

The Water Cycle

The transport and distribution of large amounts of liquid water—including its constant switching among its solid, liquid, and gaseous states—is one of the most important features of Earth's climate. When Leonardo da Vinci said, "Water is the driver of nature," he was acknowledging that water, propelled by the varying temperatures and pressures in Earth's atmosphere, allows life as we know it to exist on our planet.

Taking a closer look at the distribution of water among its various reservoirs is enlightening. More than 97% of Earth's water is in the oceans; nearly all of the remainder (the fresh water) is on the continents, predominantly in the polar ice caps. Next in volume, despite recent "mining" at far greater than replenishment rates, are the underground water reservoirs, aquifers, which contain much more water than exists in rivers and lakes. The atmosphere commonly holds only about one hundred thousandth of Earth's available water, and the biosphere still less. All the same, water constitutes about half of the weight of all living things and acts as the transport medium for essential nutrients and waste products.

From our perspective as residents on Earth's surface, it is difficult to comprehend that the oceans and atmosphere are extremely thin layers on a very thick Earth, rather like the outer layers of an onion with a 6380 km radius. A hypsometric curve plotted for Earth, showing the proportions of the surface lying at various distances above and below sea level, reveals that virtually the en-

Sandro Botticelli, The Birth of Venus, *c. 1486. The Uffizi Galleries, Florence, Italy. Throughout history wind, water and sunshine have been closely associated with the origin and maintenance of life.*

Reservoirs of Available Water on Earth

Reservoir	Volume (10⁶ km³)	Percentage of total
Oceans	1350	97.3
Glaciers (liquid equivalent)	29	2.1
Aquifers	8	0.6
Lakes and rivers	0.1	——
Soil moisture	0.1	——
Atmosphere (liquid equivalent)	0.013	——
Living biosphere (liquid equivalent)	0.001	——

flow is called the thermohaline circulation, because it arises from density variations, and water is denser when colder and/or more saline (*hals* is an ancient word for salt). High-density water sinks rather deep and, once its equilibrium depth has been reached, moves more or less horizontally. This migration of cold, salty water a few thousand meters beneath the ocean surface is balanced by compensating currents closer to the surface moving toward the sites where the water is sinking.

The combination of the thermohaline and deep circulation produces a transport system that carries water, heat, salt, and other chemicals nearly around the whole Earth. The system, shown on the following page, can be pictured as beginning in the North Atlantic, where cold, dry Arctic air removes heat and promotes evaporation, thus increasing ocean salinity (and thus density) and causing the water to sink. This water, which then flows south

tire ocean depth and mountain and tropospheric height occur within about one half of 1% of the planet's radius. Some particularly interesting characteristics of this curve are the small amount of surface that is either extremely high or extremely deep, the rather large area of the underwater continental shelf, and the relatively small differences in the heights and depths of the continents and the ocean floor.

The water in the oceans is constantly in motion, and this movement is crucial to climate because it transports heat. Ocean surface currents—those in the top few tens of meters—are well mapped and named, thanks to the tremendous importance of navigation in the exploratory, economic, and political activities of the last five centuries. These currents tend to follow the winds above them; thus, their flows approximately resemble those of the winds in the lower atmosphere, except where diverted by the presence of continents. A well-known example of heat transport by ocean currents is the Gulf Stream, which flows from Florida to northern Europe, rendering much of Europe several degrees warmer than it would otherwise be.

Ocean water is in motion far below the surface as well as near it, but the patterns are quite different. The deeper

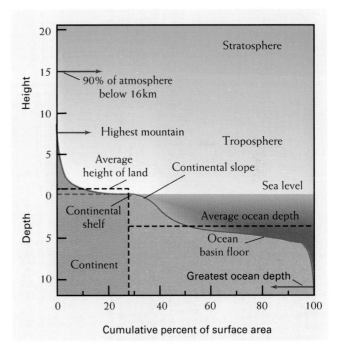

The cumulative hypsographic curve of the Earth and atmosphere; 90% of the atmospheric gases are found below 16 km altitude and nearly all the ocean water is within 7 km of the surface.

at a few thousand meters depth, forms the Atlantic deep-water current; its magnitude is more than 20 times the flows of all the world's rivers. The current turns east, passing Africa and Australia, and then travels northward. In the North Pacific, it incorporates heat and freshwater transferred from the Asian continent. Now relatively warm and less saline, it works its way back to the North Atlantic at depths of a few hundred meters.

Far different from the relatively rapid flow of the surface and thermohaline currents is the stately pace of the deepest ocean waters, which mix over hundreds or thousands of years. Their unhurried motion guarantees that, once modified, the chemistry of Earth's ocean waters will remain altered for a much longer period of time than will the chemistry of the atmosphere. This persistence is dependent, of course, on the long-term stability of the large-scale salt transport system and of the deep circulation. In Chapter 6 we will present arguments suggesting that such stability is not assured.

In the turbulent, wind-stirred upper layers of the ocean, gases, particles, and heat absorbed on the surface are mixed to about 100 m depth in a year's time. This upper-layer vertical diffusion, a random process distinct from thermohaline flows, plays an important role in moderating seasonal atmospheric temperature fluctuations. The large capacity of the oceans to store heat for long periods through this mechanism stands in marked contrast to the more limited and transitory heat capacity of the continents, where only the upper few meters exchange heat with the atmosphere.

The waters of the oceans are connected to the planet's other water reservoirs—freshwater, atmosphere, cryos-

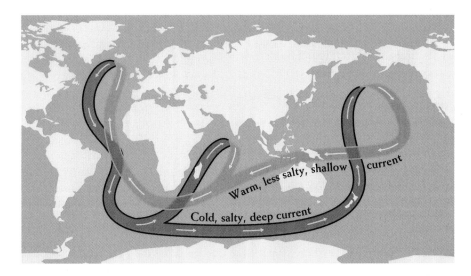

The large-scale salt transport system that operates in today's oceans. Salty, deep water formed in the North Atlantic flows down the length of the Atlantic, around Africa, through the Indian Ocean, and finally northward in the deep Pacific Ocean. This water upwells in the North Pacific, eventually working its way near the ocean surface back to the Atlantic.

Clouds are not only strikingly beautiful, they also play major roles in the radiation, water, and chemical budgets of the atmosphere.

phere—through the processes of the hydrologic cycle. The cycle can be thought of as being inaugurated by evaporation from land and sea, maintained by condensation of the evaporated water into clouds, and completed by the return of the water to the surface through various types of precipitation. The hydrologic cycle is closely related to patterns of atmospheric circulation and temperature. Changes in any of these variables can be expected to cause changes in climate and biospheric conditions.

As the Sun determines the structure and motions of the atmosphere, so too, through the mediation of the atmosphere, does the Sun drive the water cycle. The rate of water evaporation is highest when the surface is warm and the air is dry; the resulting water vapor is quickly carried aloft. Evaporation is most likely in the warmer months and during the day. Most of the water that evaporates into the atmosphere comes from the oceans, but contributions are also made by lakes and rivers, moist ground, and vegetation (the process of evaporation of water from moist plant surfaces and from the soil is termed evapotranspiration).

Water, of course, is present in the atmosphere both as vapor and as liquid droplets. It can be as abundant as a few percent of all atmospheric molecules at low altitudes

in the warm tropics and as scarce as a few parts per million in the cold lower stratosphere. In the vapor phase, water has an average lifetime of about 10 days in the atmosphere and can move thousands of kilometers before condensing. This movement occurs through two principal processes: exchanges across latitude bands (meridional moisture exchange) and exchanges between land and sea. In meridional moisture exchange, water evaporated from the subtropical ocean waters into the dry air that has descended from aloft is transported to the equator by the dominant surface winds (the Hadley circulation discussed earlier). Near the equator, the surface winds of the northern and southern hemispheres meet, causing strong upward convection of moist air, cloud formation, and precipitation.

Warm air can hold much more water vapor than cold air, so when a water-vapor-containing but unsaturated air parcel is cooled, it eventually reaches the point of saturation. This is the dew point, the juncture at which any further cooling results in the deposition of water vapor onto convenient surfaces, generally the very small atmospheric particles that serve as condensation nuclei. The droplets formed by this condensation process grow by further water vapor accretion to form clouds.

The frequency of cloud occurrence is of obvious importance to weather and climate. Clouds appear to be so substantial, and block sunlight so well, that it is truly surprising how little water they contain. Whereas the density of liquid water is 10^6 grams per cubic meter (g m^{-3}), the water content of a typical cloud is only about 0.5 g m^{-3}. However, cloud water, contained in droplets with an average diameter of one or two tens of micrometers, has about 5% of the surface area of liquid water with less than a millionth of the mass. This large surface area explains the effectiveness of cloud droplet absorption and reflection of solar radiation as well as the efficiency of cloud interactions with atmospheric particles and gases.

As air parcels move through clouds, the chemical constituents within the parcels have the opportunity to interact with the droplets; thus, the soluble constituents in the air are frequently transferred into the liquid phase. In the lower half of the troposphere, air comes in contact with clouds about 15% of the time. From measurements of cloud coverage, their typical heights and lifetimes, and the updraft velocities of air through clouds, scientists calculate that a typical air parcel is in contact with clouds for a few hours or less at a time, followed by cloud-free periods that last 5 to 20 times longer. As we will see in the following chapter, the chemistry within clouds is much different from that in cloud-free air.

In the tropics, precipitation is usually caused by water vapor condensing in rising air masses and the cloud droplets colliding and coalescing to produce progressively larger droplets. Finally a few of the hydrometeors become so large that their tendency to settle overcomes the updrafts and frictional forces holding them aloft, and they begin their fall to the ground. As they fall, they collect many smaller droplets and continue to enlarge. In cooler regions, such as the midlatitudes, precipitation often begins with the freezing of a few cloud droplets. Because water's equilibrium vapor pressure is substantially lower over ice than it is over water, the ice particles in the cloud grow faster than the water droplets, both by the deposition of water molecules on them from the gas phase and by the transfer of water molecules to them from the water droplets. This stage can easily be recognized in clouds by the diffuse, veil-like boundary that replaces the bright regions at the cloud tops, characteristic of the presence of ice crystals. When this stage in the cloud-forming process is reached, precipitation normally follows.

It was pointed out earlier that the budget for solar radiative energy is, and must be, in balance, and we infer that the water budget must adjust itself to equate inputs

Water Flow on the Continents

Continent	Precipitation rate[a]	Precipitation flux[b]	Evaporation rate[a]	Runoff rate[a]	Evaporation rate/ Precipitation rate
Africa	20.8	0.69	16.6	4.2	0.8
Asia	32.1	0.73	19.2	12.8	0.6
Australia	6.4	0.83	4.4	2.0	0.7
Europe	7.2	0.68	4.2	3.0	0.6
North America	13.9	0.57	8.0	5.9	0.6
South America	29.4	1.65	19.0	10.4	0.6

[a] 10^3 cubic kilometers of water per year.

[b] Cubic meters of water per square meter of land per year.

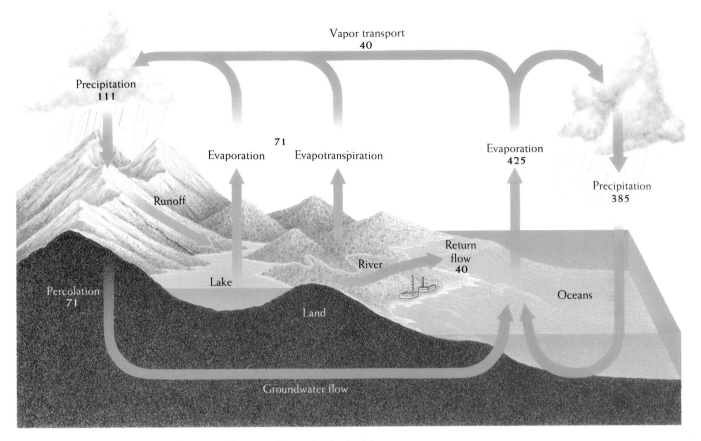

Earth's water budget. The units of the water flows are thousands of cubic kilometers per year.

and outputs as well, since, as with all budgets, such an equilibrium on a global basis is required if the water cycle and related climate are to be stable. We do not have precise information for all parts of the cycle (there are few measurements of precipitation over the oceans, for example, and oceans are 70% of the planetary surface), but we can attempt to build a water budget from the information that we do have. Some of the relevant statistics for the continents are given in the table on the facing page. When the flows on each continent are expressed in terms of area, the rates of precipitation are surprisingly uniform for most of the continental land masses. (This does not mean, however, that all areas of the continents receive similar

amounts of precipitation, simply that the *average* rainfall on each continent is similar.) The exception is South America, which receives twice the precipitation per unit area of any of the other continents. Another interesting feature is the ratio of evaporative loss to that of gain by precipitation. The fraction is about 0.6 for Asia, Europe, and both the Americas; but it is higher for Africa and Australia because of the large stretches of desert that those continents contain and the high ambient temperatures that exist there during much of the year. The land–sea moisture exchanges are in balance even though the sea loses more water by evaporation than it receives in precipitation and the land receives more by precipitation than it loses by evaporation. The balance must thus be maintained by river and groundwater flows to the oceans.

When all of the transfers and flows of water are considered on a global basis, the result is the global water budget pictured in the illustration on the previous page. Its major pathways—precipitation, evaporation, and vapor transport—are all highly sensitive to temperature, and therefore differ with geographical location, season, and climate. The oceans are primary factors in each of these pathways, receiving most of the precipitation and experiencing most of the evaporation.

A basic element of the global water budget, and a crucial factor for life on land, is the exchange of water between the oceans and the continents. More than 40,000 cubic kilometers (km^3) of water per year are transferred through the air from sea to land. The land retains part of this amount in the form of groundwater, and part eventually runs back to the sea in rivers. In the interim, it may be retained in reservoirs. Although freshwater reservoirs hold hundreds of times less water than do the oceans, much of it is at far shallower depths, so that its recycling over land is quite efficient: two-thirds of the precipitation falling on land is the result of evaporation from freshwater reservoirs.

The transpiration of water from vegetation plays a vital, if seldom appreciated, role in the cycling of water over the continents, especially in the tropical forests. Precipitation over the continents amounts to about 111,000 km^3 per year, some three times the amount of water transported from the oceans. The difference consists mostly of water that enters the atmosphere through evapotranspiration. The continental water supply is thus recycled about twice. The flows of water, and, indeed, the entire water cycle, are directly related to global temperatures and the atmospheric and oceanic circulation and have the potential to change in response to changes in those driving forces.

Extremes of Climate

As we said earlier, climate in a given location is generally characterized by the average long-term behavior of temperature, precipitation, wind flow, and the like. But another aspect of climate—the one that makes headlines—is its extremes: the 100-year flood, the blizzard of the decade, a sudden clustering of hurricanes or tornadoes. Atmospheric scientists describe most of these anomalies

The aftermath of the "Blizzard of 1888" in New York City, photographed on Sunday, March 12, 1888.

Chapter Two

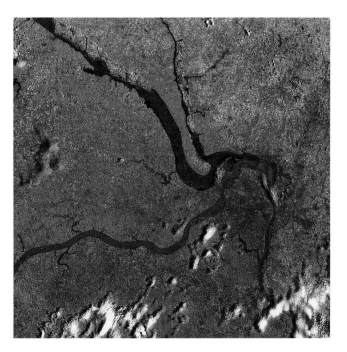

Heavy rains in the summer of 1993 produced floods along most of the Mississippi River in the central United States, as shown in these Earth satellite photographs of St. Louis, Missouri on July 4, 1988 (left) and July 18, 1993 (right). Extreme climatic events may be increasing in frequency as a consequence of added radiative absorbing gases in the atmosphere.

as random chaotic events, relatively unpredictable and relatively rare. In some cases, however, significant temporary changes in climate have direct and identifiable causes. A well-known example is the monsoons that are so much a part of the climates of eastern and southern Asia and parts of Africa. In those regions, where large continental size and high mountains accentuate many of the features of air motions, the general flow of air during most of the winter months in the lower troposphere is seaward, because the continents become cooler than the seas as solar radiation fluxes are decreased. This pattern typically creates a dry season; in contrast, the large land mass readily heats up during the summer months, bringing water-laden air flows from the oceans over the land and depositing large amounts of precipitation upon it.

More localized events such as floods and tropical cyclones are harder to explain and predict; even more difficult is predicting the frequencies with which extreme events are likely to recur. The primary obstacle to understanding any changes in the frequencies of these inherently variable phenomena is the difficulty of determining a baseline from which to calculate variations over the last few decades or centuries. A few attempts have been made to find a pattern in the extreme climatic events of the past several decades, for which the data are the most abundant and dependable. To extract clear evidence of *anything* from such a highly variable data base, however, is never an easy task, and thus far no clear evidence for an increasing incidence of extreme events has emerged. Nonetheless, any shift that does occur in Earth's climate seems likely to change as well the frequencies of extreme events such as floods, droughts, and hurricanes. In fact, one of the predictions of global warming models is that the area of ocean water warm enough to spawn hurricanes (about 27°C) will expand, producing longer and more intense storm seasons. The insurance industry is deeply troubled by the possibility of a major increase in losses from more frequent natural disasters related to climate change.

The Atmosphere as Transporter

The atmospheric circulation, the occurrence of clouds, and the flows of water on the planet are of interest to us for more than their function as creators of climate; they are also the factors that transport particles and toxic or reactive trace gases from one place in the atmosphere to another. This crucial role determines how rapidly a city is cleansed of smog once it is formed, how far dust storms travel, and how rapidly and efficiently greenhouse gases reach the upper troposphere and lower stratosphere.

Consider an air parcel that is strongly heated at Earth's surface during daytime by the Sun. Such superheated air parcels near the ground will rise, expand, and cool until they reach the same temperature and density as the surrounding air. Anyone who has flown in an airplane will have noticed the turbulence often present near the ground and lack of turbulence as the plane flies along the customary cruising altitudes in the upper troposphere and lower stratosphere. The layer within which air parcels are rapidly mixed, often about 1 to 2 km thick during the day, is termed the planetary boundary layer.

During daytime, the boundary layer can often be easily identified visually from high mountains or aircraft, because that layer tends to be filled with pollutants from various anthropogenic activities and is therefore less transparent than the rest of the atmosphere. During nighttime, the radiative cooling of Earth's surface minimizes vertical temperature differences and creates very stable atmospheric conditions. An air parcel at the surface cannot move upward, because it is cooler than the surrounding air. Under such conditions, hardly any vertical mixing of boundary layer air is possible, and intense buildups of air pollutants can occur close to the surface. A temperature

structure warmer aloft than at the ground is called an inversion. During winter, when daytime heating of the surface is small and nocturnal cooling large, inversions can be very persistent, especially if the ground is covered with snow.

Above the boundary layer, the temperature of the troposphere generally decreases with altitude. The stratosphere, in contrast, has temperatures that are either constant or increasing with height, thus producing a permanent temperature inversion above the tropopause (the troposphere–stratosphere boundary). As a result, material injected directly into the stratosphere (by aircraft emissions, volcanic eruptions, and so on) or occurring as the products of chemical reactions, such as ozone, will change altitude very slowly and can remain in the region for months or years.

In the tropical regions, the strong upward convection of hot and humid boundary layer air causes the tropopause in those latitudes to be high (16 to 18 km) and very cold (about −80°C). Because of efficient loss of water to tropical precipitation, the air entering the stratosphere there is very dry, typically with a water vapor mixing ratio of only about 3 parts per million by volume (ppmv), about 0.01% of its initial content near the surface. As a result, the stratosphere is largely devoid of clouds. It is also very clean, because the air masses that enter the stratosphere from the tropics have encountered strong condensation and freezing processes during their ascent, and so any soluble gases and particulate matter they may have contained have been efficiently removed by being incorporated into cloud drops and removed by precipitation. (These chemical species include many acids, ammonia, seasalt, and other airborne particles.) However, a number of natural and/or industrially produced gases—methane (CH_4), nitrous oxide (N_2O), and the chlorofluorocarbons (CFCs), for example—are so insoluble in water and chemically stable in the troposphere that they survive and reach the stratosphere, where they are broken down by the Sun's intense radiation and initiate many of the chemical reactions that affect ozone and that are discussed later in this book. The process, on average, is not very rapid, occurring over periods of several years, at rates dictated by the extremely inhibited upward transport of air across the tropopause and throughout the stratosphere. In fact, most

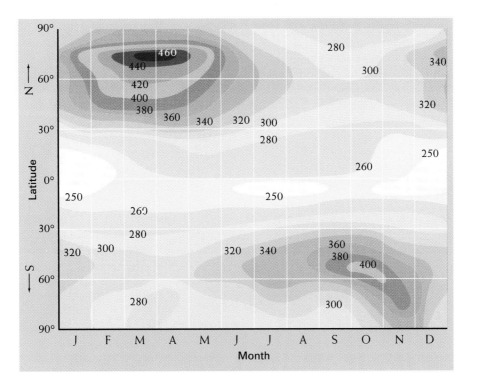

Column abundances of ozone in the unperturbed stratosphere (i.e., the abundance measured within a column from the planetary surface to the top of the atmosphere) as a function of latitude and season. The abundances are expressed in Dobson units: 1 DU = 2.69 × 10^{16} molecules of O_3 cm^{-2}; 100 Dobson units corresponds to an ozone layer thickness of 1 millimeter if all ozone in a vertical column were at 1 atmosphere pressure and 0°C.

air exchange across the tropopause is probably caused not by slow diffusive mixing but by uncommonly vigorous convective thunderstorms that rapidly inject tropical tropospheric air into the stratosphere.

The altitude of the tropopause in the tropics is 6 to 9 km higher than its altitude in higher latitudes. This has a significant effect on the distribution of the total ozone column, the amount of ozone in a column extending from Earth's surface to space. Most ozone is created by photochemical reactions in the tropical stratosphere at altitudes above 25 km; it is then transported to the lower stratosphere in the higher latitudes and accumulated there due to slow downward mixing. The result is that the lowest total ozone occurs in the tropics, as seen in the diagram on this page; consequently, by far the most ultraviolet radiation penetrates to Earth's surface in the tropical regions. In other words, the tropical biosphere has historically been poorly protected from solar ultraviolet radiation, and has of necessity developed the strongest defense mechanisms against it (an example is the dark skin color of tropical natives).

Those who study atmospheric chemistry are particularly interested in the velocities of movement of airborne species between different atmospheric regions. For example, they know that the horizontal flow in the lower tro-posphere, but above the boundary layer, moves at veloci-ties of the order of 5 m s^{-1}. This velocity implies that emissions from automobiles or industrial smokestacks will move several hundred kilometers from the point of origin within a day or two, provided they are not deposited on nearby surfaces or rapidly transformed by chemical reac-tions. Species whose limited reactivity or solubility allows them to remain in the atmosphere as long as or longer than a few days are therefore effectively moved from their place of emission to locations across national boundaries. A familiar example is that of the sulfur and nitrogen ox-ide emissions manifested in acid rain, with continental Europe and the United States exporting the pollution by air to Sweden and Canada, respectively. Another isolated example was the atmospheric transport of radioactive sub-stances from the 1986 Chernobyl nuclear power plant ac-cident in the Ukraine to various parts of Europe. In north-ern Scandinavia the deposition of radioactive debris was so heavy that large portions of the reindeer herds of the Lappish people had to be slaughtered.

At higher altitudes, but still within the troposphere, the strong zonal winds can move trace atmospheric con-stituents with long lifetimes across large portions of the globe in a week or two. It is these winds that transport dust from the deserts of China to the far different land-scapes of Hawaii or Alaska, from the Sahara Desert to Florida or Europe, and from central Russia to the Arctic. If trace species are directly injected into the atmosphere at these altitudes, their transport can render their effects widespread. An example of this potential was provided in 1991 by the eruption of Mt. Pinatubo in the Philippines.

Emissions from a steel foundry in Gary, Indiana, in 1950, before emission controls were installed. The average wind flow will move these emissions, or their chemical products, to the East Coast of the United States (a distance of some 700 km) in a day or so.

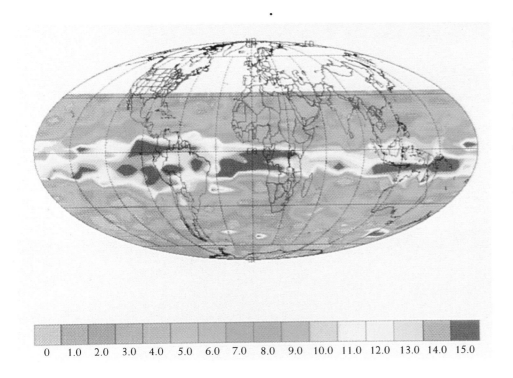

Stratospheric sulfur dioxide as injected by the Mt. Pinatubo, Philippines, volcano in June, 1991, and as detected on September 19, 1991, by the Upper Atmosphere Research Satellite. The color bar units are parts per million by volume (ppmv). From its eruption site near the equator in Asia, the gas has clearly been spread throughout all longitudes by the stratospheric winds.

| 0 | 1.0 | 2.0 | 3.0 | 4.0 | 5.0 | 6.0 | 7.0 | 8.0 | 9.0 | 10.0 | 11.0 | 12.0 | 13.0 | 14.0 | 15.0 |

This large volcano shot substantial amounts of dust and gases into the upper troposphere and stratosphere. Upper atmospheric motions then carried that debris eastward at an average zonal speed of approximately 25 m s^{-1}, circling Earth in an undulating manner every 15 to 16 days and eventually causing worldwide cooling following its global mixing throughout the stratosphere, a process that took about half a year.

Overview

The ideas explored in this chapter will serve as building blocks for our discussions throughout the rest of the book. The most basic concept is that weather and climate arise as a consequence of solar radiation in the atmosphere in combination with the rotation of Earth. These forces produce a general atmospheric circulation with reasonably predictable behavior both in the vertical and horizontal directions—circulation that transports energy, water, and pollutants over long distances. In fact, climate can be defined as the long-term average of energy and water flows in the atmosphere and on the land, for Earth as a whole or at a specific geographical location. A perspective on Earth's interactions with radiant energy and with water is provided by constructing budgets for each. The budgets must be in balance; that is, Earth as a whole is neither gaining nor losing radiant energy nor gaining nor losing water, although we cannot necessarily measure each input and output factor accurately. A change in any input or output will require a compensating change in one or more of the other inputs or outputs, or else a new level of stability will result.

The relative stability of climate is obviously vital for such important activities as human settlement, agriculture, and trade. We will discuss in later chapters what is known concerning the stability of climate in the recent and more distant past, and what we might predict for climates of the future.

Final inflation of a scientific balloon before launch at Europe's Kiruna, Sweden facility. The balloons are made of thin polyethylene material, and provide atmospheric measurements up to 50 km in altitude.

Chemistry in the Air | 3

There once was a scientist named Newton
Whose equations were not highfalutin.
They governed—though terse—
The entire universe.
And further, they did no pollutin'.
—Max Shulman

*C*hemistry is most often thought of as something that happens in liquids, especially the highly colored liquids that are a feature of science classes and science fiction movies. But the full range of chemistry—slow and fast reactions, dissolving crystals, precipitation of colored solids—occurs in the atmosphere as well. In fact, measurements taken on land and from aircraft, balloons, and satellites have revealed that the atmosphere, which we might term a "flask without walls," contains several thousand different chemical species. Some are components of the atmospheric gas itself, some are found in airborne particles, both large and small, some are found dissolved in hydrometeors. A number of them appear in more than one, even all, of these reservoirs. Many atmospheric compounds have both natural and anthropogenic origins and thus were present in the atmosphere before the Industrial Revolution as well as today.

More than 99.9% of the molecules constituting Earth's atmosphere are nitrogen, oxygen, and chemically inert "noble gases" (largely argon), if the highly variable amounts of water vapor are disregarded. This 99.9% has been present and chemically stable at nearly constant levels for the past several hundred million years. The same cannot be said of many of the other species, whose interactions are so common that they influence our everyday lives. In retrospect, it should not have been surprising to discover this chemistry, since transformation reactions in oceans, for example, have long been accepted facts. Atmospheric chemistry presented greater logistical and conceptual challenges, however—the reactions were more rapid, the dominant processes unknown, and the concentrations of the interacting trace species minute. And once again, as in so much of atmospheric science, the key question concerning those trace species is related to budgets: Even though these atmospheric constituents are formed and removed rapidly, is their overall influence stable over time?

Of the chemically reactive compounds, methane is by far the most abundant, at a current northern hemisphere ground-level concentration of about 1.7 parts per million by volume (ppmv). In fact, the total amount of all reactive molecules present in the atmosphere seldom exceeds 10 ppmv anywhere in the world at any time. It is astonishing, but easily demonstrated, that all the atmospheric problems currently threatening the delicate balance of life on Earth—smog, the ozone hole, acid rain, and so on—are a consequence of chemical interactions among less than one thousandth of one percent of all of the molecules in the atmosphere.

Because of the great variability from place to place of emissions and reactions, there is no typical trace species composition for a given volume of air. Nonetheless, of the several thousand trace constituents that have been detected in tropospheric air, about 170 distinct interacting chemical species, many of them of natural origin, are commonly present in the air at any location near the surface of the Earth. In the stratosphere, the simultaneous presence of perhaps 40 reactive species is typical. The task of the atmospheric chemist is to understand at least the most influential of the reactions linking these chemical constituents, as well as their principal sources and removal mechanisms. We begin our discussion with the stratosphere, the chemically simpler of the two regimes.

Stratospheric Chemistry

Ozone is the central species in the stratosphere's chemistry, and about 90% of all atmospheric ozone is located there. Nonetheless, the stratosphere contains, at most,

"This isn't going to do the old ozone layer any good, that's for sure."

Nearly everyone has heard of atmospheric ozone, but few people understand the factors responsible for its formation and decay. Here's one you may not have thought of. (©1987, The New Yorker Magazine, Inc.)

only about 10 ozone molecules per million molecules of air. In spite of its scarcity, ozone (O_3) readily interacts with many stratospheric constituents. That, along with its strong absorption of biologically harmful ultraviolet radiation, ensures its central role.

In the previous chapter we showed how O_3 is produced in the stratosphere when shortwave ultraviolet radiation causes molecular oxygen (O_2) to split into two oxygen atoms. Although this photodissociation is an ongoing activity, O_2 remains abundant; hence, it is clear that there must also be processes that reconvert O_3 into O_2. The initial hypothesis proposed by pioneer atmospheric scientist Sidney Chapman of the University of Oxford in 1930 was that this recycling process occurred mainly through a pair of reactions:

$$O_3 + h\nu \rightarrow O + O_2 \qquad (3.1)$$
$$O + O_3 \rightarrow 2O_2 \qquad (3.2)$$

where $h\nu$ is the symbol for a single photon of appropriate wavelength. The net result of this sequence is that two O_3 molecules are converted into three O_2 molecules.

For about 40 years, it was generally accepted that these two reactions completed the cycle of stratospheric ozone.

However, research over the past two decades has shown that several minor atmospheric constituents (symbolized below as X· and XO·) also play essential roles in stratospheric ozone destruction. Their participation in the ozone reaction cycles may be summarized as

$$X\cdot + O_3 \rightarrow XO\cdot + O_2 \qquad (3.3)$$

$$O_3 + h\nu \rightarrow O + O_2 \qquad (3.1)$$

$$O + XO\cdot \rightarrow X\cdot + O_2 \qquad (3.4)$$

The net outcome of the catalytic cycle portrayed in reactions 3.3, 3.1, and 3.4 is that two ozone molecules are converted into three oxygen molecules, with one photon providing the necessary energy.

A catalytic cycle is a sequence of reactions in which at least one reactant (a catalyst) is regenerated to repeat its actions. Thus, catalysts can remove many, many more target molecules than if they were immediately consumed. The atmospheric catalysts X· and XO· are free radicals (that is, molecular fragments). Following the usual convention of atmospheric chemists, we indicate free radical species with a centered dot to signify that they have an odd number of electrons and are thus highly reactive.

Why should fragments of molecules exist in the atmosphere? As with so many other atmospheric processes, the cause is radiation from the Sun. An example of the Sun's ability to produce free radicals involves the ever-important ozone molecule. In Chapter 2 we saw that ozone readily dissociates upon absorbing solar radiation. A small fraction of that radiation has enough energy to produce not just a free oxygen atom, but a highly excited one, which we designate as O^*. This atom is so energetic that it is capable of attaching to and cleaving a water vapor molecule to generate two hydroxyl radicals ($2\,HO\cdot$):

$$H_2O + O^* \rightarrow 2HO\cdot \qquad (3.5)$$

The result is the inauguration of a catalytic ozone destruction cycle involving HO· and HO2· (hydroperoxyl) in which HO· becomes the X· of equations 3.3 and 3.4, so that the sequence is

$$HO\cdot + O_3 \rightarrow HO_2\cdot + O_2 \qquad (3.6)$$

$$O_3 + h\nu \rightarrow O + O_2 \qquad (3.1)$$

$$O + HO_2\cdot \rightarrow HO\cdot + O_2 \qquad (3.7)$$

Despite the very low water vapor content of the stratosphere, enough of these radical species are generated to affect the ozone budget of the upper stratosphere.

A second catalytic cycle, the most important one under natural conditions, was proposed by Paul Crutzen (then at Oxford University) in 1970, while he was studying the possible involvement of nitric oxide (NO) and nitrogen dioxide (NO2) in the chemistry of the stratosphere[‡]. Most of the NO_x (the general symbol for the sum of NO and NO2, popularly pronounced "nox") in the stratosphere is a consequence of ground-level emissions of nitrous oxide (N2O), largely from microbiological processes in soils and oceans. N2O is unreactive in the troposphere. Eventually, however, atmospheric motions bring it into the stratosphere, where its interactions with ozone are initiated, as with the HO_x (HO· plus HO2·) cycle mentioned above, by reaction with O^*:

$$N_2O + O^* \rightarrow 2NO \qquad (3.8)$$

The release of NO initiates the catalytic ozone destruction cycle of reactions 3.3, 3.1, and 3.4, with X· = NO.

Nitrous oxide oxidation is not the only source of stratospheric NO, though it is the most important one. Nitric oxide can also be injected directly into the stratosphere by high-flying aircraft, for example. Negligible amounts result from the transport of NO_x emitted by automobiles and power plants at ground level, however, because that material is efficiently converted to the highly water-soluble nitric acid (HNO3) and removed from the air by rainfall before atmospheric motions can move it very far from its origins.

A third catalytic cycle for ozone destruction in the stratosphere involves chlorine (Cl). It was proposed in 1974 by Richard Stolarski and Ralph Cicerone, then of the University of Michigan, who had been looking for possible atmospheric consequences of emissions from satellite rocket engines. The combustion of a common rocket fuel, ammonium perchlorate (NH_4ClO_4), spews hydrochloric

‡ NO and NO2 are technically free radicals, since they have unpaired electrons, but they are so much more stable and longer lived than most free radicals that they represent a sort of intermediate class between radicals and molecules. The usual convention is not to use the free radical dot notation for them. We follow that convention here.

acid (HCl) into a "tunnel" of the atmosphere behind the rocket, and Stolarski and Cicerone were concerned about ozone-destroying reactions in that region. Their analysis indicated, however, that the amount of HCl emitted was too small to be important. Nonetheless, they had introduced chlorine chemistry into stratospheric science, and their work eventually led to a realization that a catalytic chlorine cycle could be inaugurated naturally by methyl chloride (CH_3Cl), a natural by-product of algal photosynthesis. Methyl chloride is mildly reactive, so more than 90% of the CH_3Cl molecules emitted into the lower atmosphere are removed there. The small fraction that survives destruction in the troposphere is eventually transported to the stratosphere, where it is broken down by reaction with $HO\cdot$:

$$CH_3Cl + HO\cdot \xrightarrow{\text{several steps}} Cl\cdot + \text{other fragments} \quad (3.9)$$

$Cl\cdot$ and its partner radical $ClO\cdot$ are thus available to participate in a catalytic reduction of ozone by reactions 3.3 to 3.4 ($X\cdot = Cl\cdot$ and $XO\cdot = ClO\cdot$).

With the three major catalytic cycles occurring simultaneously—and all the different catalysts reacting with each other, besides—an exact analysis of the situation becomes a complex exercise in competitive reaction chemistry. Because all the cycles occur naturally, at least to some degree, we can assume that they have been a part of the chemistry of the air since long before humans walked the pedosphere. But can these catalytic cycles continue forever? The answer is no, they cannot. They are limited by a natural constraint: mutual self-destruction, as shown in this pair of reactions:

$$HO\cdot + NO_2 \rightarrow HNO_3 \quad (3.10)$$
$$ClO\cdot + NO_2 \rightarrow ClONO_2 \quad (3.11)$$

HNO_3 (nitric acid) and $ClONO_2$ (chlorine nitrate) are termed "reservoir molecules," since they function as temporary holders of catalytic radicals in forms unreactive to ozone. The reservoir molecules can eventually relinquish the catalytic radical fragments, as a consequence of energetic reactions that regenerate the $X\cdot$ and $XO\cdot$ species, but the susceptibility of these catalysts to reservoir mole-

cule formation serves as an injection of molasses into an otherwise vigorous chemical system.

How do we know that these reactions involving transitory chemical intermediates really occur in the stratosphere, at altitudes well beyond the reach of instrument-carrying aircraft? There are several ways (for an example, see page 40), but often the most successful has been to lift measuring equipment to the proper altitude in helium-filled balloons. In this specialty, an art as much as a science, the acknowledged champion is James Anderson of Harvard University, who has flown increasingly sophisticated detectors from balloons since 1975.

At one time or another, Anderson's instruments have detected the presence of most of the catalytic species involved in the three ozone destruction cycles: O, $HO\cdot$, $HO_2\cdot$, $ClO\cdot$, $Cl\cdot$, O_3, and N_2O. A typical detector in one of his flight packages is designed to measure, with considerable accuracy, one reactant molecule in ten thousand million molecules of inert atmospheric gas, a dilution similar to that of a drop of vermouth in a railroad tank car of gin. Anderson's technique is to allow stratospheric air to flow through a chamber of the detecting apparatus. If the molecule of interest is not conveniently detectable in itself, he may inject a precise amount of additional gas into the airflow, one with which the molecule will react to produce a more detectable species. In any case, the samples of stratospheric air are irradiated by a carefully selected stream of ultraviolet radiation, which is absorbed by the constituent Anderson wishes to measure. After a tiny fraction of a second, the absorbing molecule re-emits radiation that distinctively signals its presence, and this radiation is detected by sensors strategically placed around the outer walls of the chamber. All of this chemical artistry is accomplished in rarefied air, under cold conditions, and with few chances for new flights should a first attempt fail.

In Anderson's early balloon flights, the instruments were rapidly lifted to high altitudes and the measurements were carried out as the balloon slowly descended. After those successes, his experiments became increasingly innovative, with one series involving instruments mounted on a winch and cable, a sort of giant yo-yo, as shown on the facing page. In those experiments, the balloon maintained its altitude long enough for the Sun angle to change, and the instrument package was raised and lowered several times. The detectors were thus able to measure the varying concentrations of transitory intermediates

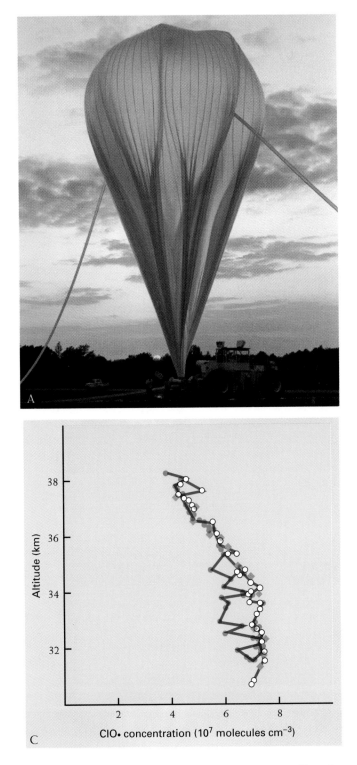

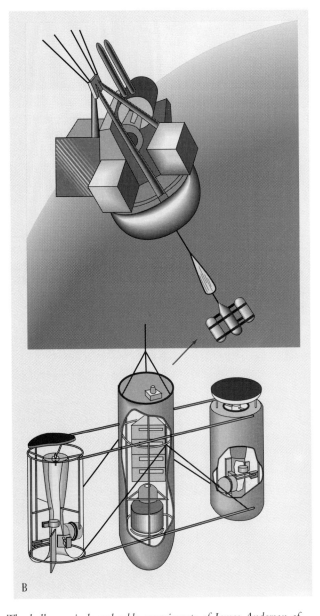

The balloon winch and cable experiments of James Anderson of Harvard University. (A) The helium-filled balloon at launch. (B) Diagram of the winch and cable package. The two pods suspended from the cable can contain duplicate or different instruments depending on the experiment design. (C) Observed concentrations of ClO over Palestine, Texas, on September 14, 1984. The yellow diamonds are averages from the instrument in Pod 1 on the first descent, the closed and open circles are averages from Pod 2 on the first and second descents.

Atmospheric Chemistry by Satellite

*T*here are many different ways of detecting atmospheric gases and particles from spacecraft, but a single example can be used to illustrate the general process. The Upper Atmosphere Research Satellite (UARS) mission was launched from the space shuttle *Discovery* on September 12, 1993, and carried several instruments, one of which, the halogen occultation experiment (HALOE), was designed to detect a number of atmospheric constituents, including those containing chlorine and bromine (the latter being members of the halogen family of elements that comprise Group VIIA of the periodic table). As illustrated on these pages, the HALOE equipment looks toward the Sun through the upper part of Earth's atmosphere, and, as the sunlight traverses the atmosphere to reach the satellite, each of the atmospheric gases absorbs distinctive portions of the solar spectrum. The radiation that has not been absorbed by the atmosphere reaches the satellite and passes through a number of carefully selected optical filters in the HALOE telescope. It then travels to a set of detectors and is measured. Each atmospheric molecule absorbs a characteristic combination of wavelengths. Therefore, a mathematical analysis of the signals from the detectors enables scientists to determine the average concentrations of those molecules along the path from the Sun through the atmosphere to the spacecraft.

HALOE observations of stratospheric nitrogen dioxide are pictured at the bottom of the facing page. This "NO_2 map" for different altitudes and latitudes was made by combining the HALOE results from many orbits over the period September 21 to October 15, 1992. The vertical scale is atmospheric pressure in millibars (the pressure at sea level is about 1000 mbar); the stratosphere is between the tropopause (approximately 150 mbar) and the stratopause (approximately 1 mbar). The display shows an NO_2 level just above the tropopause and just below the stratopause that is below 1 part per billion by volume (ppbv) at all latitudes. A strong maximum in concentration occurs in the middle stratosphere, a consequence of the nitrous oxide production of NO by reaction 3.8, followed by the reaction of NO with ozone to produce NO_2.

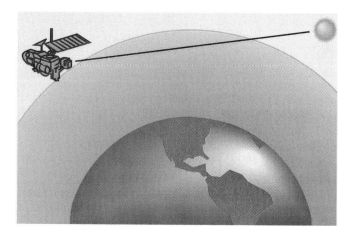

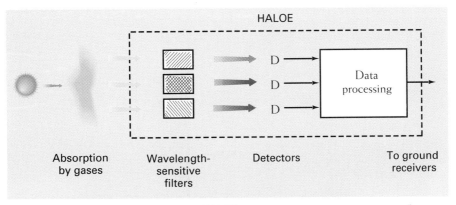

HALOE

Absorption
by gases

Wavelength-
sensitive
filters

Detectors

To ground
receivers

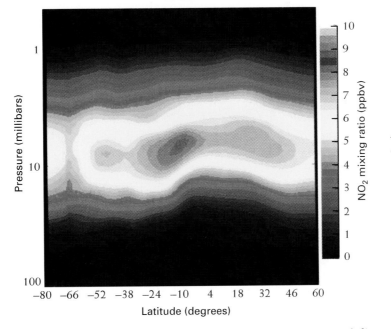

HALOE

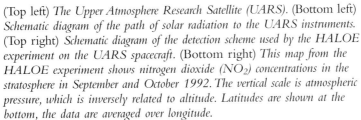

(Top left) *The Upper Atmosphere Research Satellite (UARS).* (Bottom left) *Schematic diagram of the path of solar radiation to the UARS instruments.* (Top right) *Schematic diagram of the detection scheme used by the HALOE experiment on the UARS spacecraft.* (Bottom right) *This map from the HALOE experiment shows nitrogen dioxide (NO_2) concentrations in the stratosphere in September and October 1992. The vertical scale is atmospheric pressure, which is inversely related to altitude. Latitudes are shown at the bottom, the data are averaged over longitude.*

as the latter evolved under the influence of varying amounts and wavelengths of solar radiation and of varying chemical conditions at different altitudes. Anderson's newest activity is the development of remotely controlled, lightweight aircraft that promise to reach the high layers of the stratosphere, conducting measurements there for hours or days.

Useful data on the presence of certain molecules in the stratosphere have also been provided by satellites, not at a single location, as balloon experiments do, but over much or all of the globe. These vehicles carry instruments into Earth orbit, allowing scientists to look *down* through the atmosphere, so to speak, and measure the total abundance of a molecule of interest from the top to the bottom of a vertical atmospheric column. Other satellite instruments look at the planet edge-on, peering *sideways* through the atmosphere to determine the concentration of a given molecule as a function of altitude (as we saw on page 40). Not all molecules are detectable in these ways, and even when they are, complications such as clouds and less-than-ideal orbits affect the accuracy with which they can be measured. Nevertheless, the information gathered by spacecraft is a valuable supplement to the more localized balloon and aircraft measurements.

Urban Smog Chemistry

To greater or lesser degrees, industrial emissions have been released throughout the more developed world for over a century, and yet the emitted substances have not been observed to accumulate in the atmosphere at rates equal to their release. It is well known that the pollution does not react with the very stable nitrogen and oxygen of the atmosphere, so where has it all gone? For large particles the explanation is simple: they are pulled to the ground by gravity. The fate of gaseous emissions is more complex, though their observed concentration patterns make it clear that the rates of removal must be comparable in magnitude to the rates at which the pollutants are produced.

Scientists suggested early on that ozone might be acting as an atmospheric scavenger, largely because it is known to be highly reactive. However, in the early part of this century, observations from the ground and from balloons showed that most of the atmosphere's ozone is located in the stratosphere, with the peak concentration occurring at altitudes between 15 and 30 km. Although ozone was detected in the troposphere as well, tropospheric ozone was thought to have originated in the stratosphere, from which it occasionally descended to lower altitudes through atmospheric motions, eventually to be destroyed by contact with Earth's surface. This belief was based on the fact that the photodissociation of molecular oxygen to produce ozone can occur only at wavelengths shorter than 240 nm. Because such short-wavelength radiation is found only in the stratosphere, no tropospheric ozone production is possible by this particular mechanism. But, in the mid-1940s, repeated heavy injury to vegetable crops in the Los Angeles area was traced to such high concentrations of ozone that it became obvious that ozone was being formed at low altitudes as well as high. Thus was revealed a dichotomy that nonscientists often find confusing: ozone in the stratosphere is basically a good thing, because it absorbs damaging ultraviolet radiation, but ozone in the urban boundary layer is basically a bad thing, because it is toxic to vegetation and humans.

The overall reaction mechanism for the generation of ozone in the boundary layer was eventually identified in the 1950s by Arie Haagen-Smit of the California Institute of Technology. It begins with the general reaction

$$RHC + NO + \text{solar radiation} \rightarrow NO_2 + \text{other products} \qquad (3.12)$$

where RHC denotes various reactive hydrocarbons (ethylene, butane, and other compounds consisting chiefly of carbon and hydrogen). Once NO_2 is formed, some of the solar radiation that penetrates through the atmosphere to the ground is energetic enough to dissociate it:

$$NO_2 + h\nu \rightarrow NO + O \qquad (3.13)$$

and the recombination of O with molecular oxygen produces ozone by the reaction introduced in Chapter 2:

$$O + O_2 + M \rightarrow O_3 + M \qquad (3.14)$$

It is important to note here a crucial difference between the stratosphere and the troposphere: in the stratosphere, NO_x ($NO + NO_2$) serves as a catalyst for ozone *destruction*; in the troposphere, it is a catalyst for ozone *production*. This difference occurs because the troposphere's reactive hydrocarbons are absent from the stratosphere (thus negating reaction 3.12 at high altitudes), and because O atoms are virtually absent from the troposphere (thus negating reaction 3.4 at low altitudes).

Reactive hydrocarbons have long been a standard feature of Earth's atmosphere, since plants produce and release them. Thus, it is often the presence or absence of NO_x that determines whether ozone will be generated in the troposphere. Before human technology progressed to the point of radically altering atmospheric chemistry, the high combustion temperatures needed to produce NO_x from atmospheric N_2 and O_2 were present only on occasion, during forest fires or lightning flashes. Now they are as common as the cylinders of the nearest automobile.

Sufficient quantities of NO_2 in the atmosphere will influence visibility as well as chemistry, for NO_2 absorbs blue photons, thus changing the apparent color of the sky. The loss of these photons is all the more noticeable if large quantities of light-scattering particles are present as well. The final result is smog, the familiar brownish haze seen in the air of sunlit, traffic-clogged cities around the world.

Because the Haagen-Smit mechanism of ozone generation requires the presence of both RHC and NO_x, the sources of both these families of compounds must be considered before an effective remedy for smog can be prescribed. President Ronald Reagan was ridiculed by the news media in the 1980s for commenting that trees caused air pollution, but he was right in part. It is true that the major sources of RHC and NO_x in cities are generally fossil fuel combustion and a variety of industrial processes. However, RHCs are also emitted by vegetation (for example, RHC molecules called terpenes are what create the attractive smell of an evergreen forest). Urban ozone generation is most efficient when the relative proportion of RHC to NO_x is about 7 to 1, and vegetative emissions are often abundant enough to have a critical influence on the ratio in a given location, especially in suburban environs near cities. In some urban-suburban areas, such as Atlanta, Georgia, the RHC emissions from vegetation are substantial enough that they must be taken into account in the design of regulatory strategies for reducing ozone.

The presence of ozone in the troposphere as a consequence of smog reactions explained part of the "where does it all go?" question, because ozone reacts with some of the

Photochemical smog obscures the ancient minarets of Ankara, Turkey, in a scene increasingly reminiscent of summer days in urban areas the world over.

RHCs to generate moderately soluble products that are eventually absorbed into raindrops and washed from the air. However, ozone is not reactive enough toward a great many trace molecules to explain their disappearance from the atmosphere. The first step in understanding the fate of those species occurred in 1971, when Hiram Levy, then of Harvard University, pointed out that ozone photodissociation by solar radiation in the troposphere could lead to the production of the hydroxyl radical just as occurs in the stratosphere (reactions 3.1 and 3.5). The hydroxyl radical (HO·) turns out to be the "principal atmospheric detergent," the reactant that scrubs most of the contaminants from the air. HO· behaves in this way because it is a fragment of the very stable water molecule, to which it wants to revert by wresting (abstracting) a hydrogen atom from another molecule nearby. As a very great percentage of the molecules in urban emissions contain one or more hydrogen atoms, HO· abstraction is a common process.

The chemical equations describing the HO· abstraction sequence for a typical molecule show how important the process is to atmospheric self-cleansing. Almost any pollutant containing a hydrogen atom could serve as an example, but methane (CH_4) is the simplest, requiring only four steps:

$$CH_4 + HO· \rightarrow CH_3· + H_2O \qquad (3.15)$$

$$CH_3· + O_2 \rightarrow CH_3O_2· \qquad (3.16)$$

$$CH_3O_2· + NO \rightarrow CH_3O· + NO_2 \qquad (3.17)$$

$$CH_3O· + O_2 \rightarrow HCHO + HO_2· \qquad (3.18)$$

Why does this reaction sequence cleanse the air? It is because once the initial HO· attack has occurred, the resulting molecular fragment ($CH_3·$ in this case) readily attaches to the abundant O_2, and oxidized molecules are generally eager to bind with H_2O. For example, the sequence above begins with CH_4 and concludes with formaldehyde (HCHO), a species about a million times more water soluble than its precursor. Thus, it can be rapidly scavenged from the atmosphere by clouds and rain. Without the cleansing activity of HO·, the current composition of the atmosphere would be totally different and perhaps hazardous to many of the present forms of life on

Earth. It is truly remarkable that such large amounts of toxic air pollutants and greenhouse gases are removed from the atmosphere primarily through the agency of a few hydroxyl radicals, typically present at concentrations of the order of a few parts in 10^{14} parts of air.

Examination of the methane reaction sequence shows that the number of reactive free radical species is conserved in each step; that is, a dot indicating the presence of an unpaired electron appears on each side of the equation. Why are we not overwhelmed by free radicals in the boundary layer? As in the stratosphere, the answer is that free radicals self-destruct. For example,

$$HO_2\cdot + HO_2\cdot \rightarrow H_2O_2 + O_2 \qquad (3.19)$$

the two free radicals reacting to form stable products. Hydrogen peroxide (H_2O_2) is one of the most soluble of all atmospheric gases and is rapidly removed by the fog over any nearby lake or river or by droplets in a passing cloud.

Thus, the principal factor influencing the atmospheric chemistry of Earth is the formation by solar energy of the oxidizers HO· and O_3, which react with and break down virtually all atmospheric trace gases. The several roles of ozone are crucial to the overall process. First, it is the species in the upper atmosphere that absorbs most of the ultraviolet radiation to which all biological life is susceptible. Second, it generates HO· in the lower atmosphere

and thus cleanses the air. In the stratosphere, any systematic loss of ozone is to be avoided. In the troposphere, ozone's toxicity and its involvement in urban smog formation argue against allowing it to be generated in a profligate manner. Nonetheless, at modest levels it plays a vital part in preserving the desirable characteristics of the atmosphere in which we live and which we breathe.

Influences Near and Far

Despite the efforts of HO· and O_3, not all pollutants are promptly removed from the air by reaction, or by deposition. Those that are not have the potential to affect climate and air quality not only in the urban areas where they are emitted, but immediately downwind of those regions too, often over very large areas. A graphic example of how city traffic and industry can affect climatic variables was provided by a study of precipitation in and around St. Louis, Missouri, for several years in the 1970s. Readings from rain gauges in the area revealed a clear pattern of higher rainfall downwind of St. Louis than upwind or within the city, as shown in the diagrams below. Obviously the city itself was not producing cloud droplets, but just as obviously the city was somehow generating them indirectly. The culprits turned out to be small particulate pollutants that acted as nuclei around which cloud droplets formed, and moisture emitted from industrial,

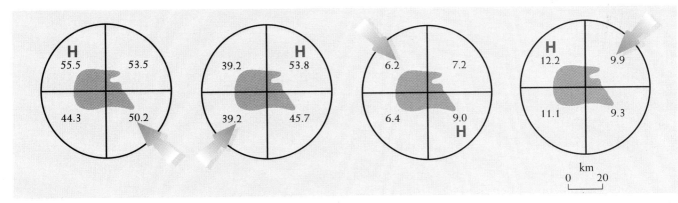

*Total summer rainfall as a function of location for the St. Louis, Missouri, metropolitan area. The metropolitan area is outlined on each plot. The plots are for four groups of rainfalls during 1971 to 1975, subdivided on the basis of the prevailing wind direction at the time of the rainstorm: (left) southeast, (second from left) southwest, (second from right) northwest, and (right) northeast. **H** indicates the quadrant in which the highest average rainfall occurred (given in millimeters).*

commercial, and residential sources, which further enhanced the formation of clouds. Among the inadvertent results of that study has been an increase in the cost of storm and hail insurance for all farmers whose land is generally downwind of the city. As compensation, the added rainfall seems to slightly increase their crop production relative to that of their upwind competitors.

We have said little thus far about airborne particles. While they, too, produce climate effects, they do so by physical means (decreasing the penetration of light, providing nuclei for cloud droplet condensation, and so forth), rather than chemical ones. The distances over which they exert their effects are related to how quickly they are removed from the atmosphere by rainfall or grav-

ity, but tend to be local or regional in scale. Emissions consisting of chemically reactive *gases*, however, can be transported many hundreds of kilometers away from the site of their emission. In 1976, William Cleveland and colleagues at Bell Laboratories described a vivid example. They had collected hour-by-hour measurements of ozone and other pollutants from 32 locations throughout the northeastern United States and made plots of the ozone data for days on which the wind flow was consistent and relatively uniform, as shown in the maps below. On July 2, 1974, that flow was from the southwest during the entire day. Therefore, anything injected into the air in the cities of Washington, Baltimore, Wilmington, and Philadelphia was blown to New York, and on to Boston, with the wind pick-

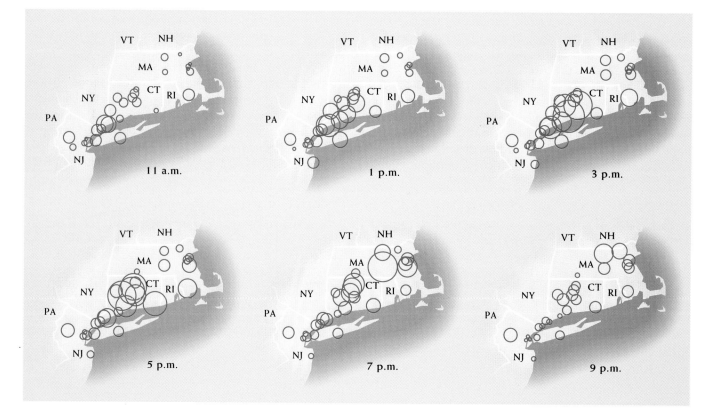

Ozone concentrations, coded by circle sizes, plotted as a function of time of day for the northeast United States, July 2, 1974. The largest circles (highest ozone concentrations) occurred sequentially over New York, Connecticut, Massachusetts, and New Hampshire, on a day when the prevailing wind was from the southwest.

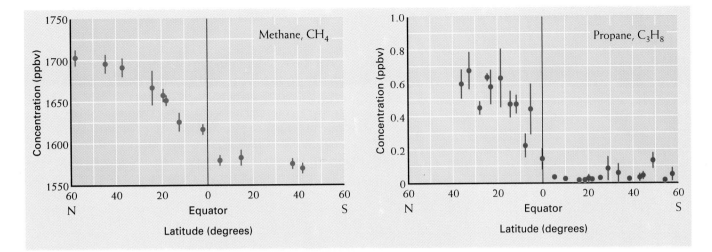

Interhemispheric profile of the surface air concentrations of methane and propane. Methane is regularly measured at a number of sites around the world, and the data shown here are from annual mean measurements at a number of sites. The data for propane were obtained from measurements aboard the U.S. Coast Guard vessel Polar Star, *which sailed from Seattle, Washington to Punta Arenas, Chile during the period from November 15 to December 28, 1984. The error bars reflect the relative noisiness of the data at each site after seasonal and secular terms have been removed.*

ing up added injected material along the way. By evening, the air that had begun the day over Washington had moved into southern New Hampshire.

Washington's automobiles and other emissions sources had begun the smog-creating process by generating NO_2 and RHC before the sun had fully risen. That air moved northeast as the day progressed and as the sun rose higher in the sky. Solar radiation began to dissociate the NO_2 and generate ozone as the air parcel moved along. At midday, high ozone levels were detected over New York. They were then seen, sequentially, in western Connecticut, eastern Connecticut, and Boston. By late evening, well after the sun had set, a level of ozone that exceeded air quality standards for human health and that had been produced in part by emissions six states upwind was detected near the forests of New Hampshire.

Air-transport effects can be seen globally as well. At the equator, for example, converging flows of air from north and south meet at the intertropical convergence zone. The bulk of these flows tends to rise, then separate, each stream moving poleward at high altitudes within the hemisphere from which it originated. Some of these streams, however, make their way *across* the equator. Thus, chemical species emitted at the ground in one hemisphere may affect the lower atmosphere of the other. Because cross-equatorial flow is slow, air exchange times between hemispheres of about a year are typical. If the gas of interest is emitted predominantly in one hemisphere and has a long lifetime, the result is a very modest concentration gradient across the latitudes, as is the case with methane in the graph above. A more reactive gas will show a more dramatic interhemispheric difference in concentrations, because it is removed more efficiently by reactions with HO· or O_3 near its source before it has time to reach the opposite hemisphere. This is the situation for propane, shown in the graph on the right. Its concentration decreases from about 0.6 ppbv in the northern midlatitudes to almost nothing across the equator. When the emissions sources are not well known for a particular species, such concentration patterns, combined with a knowledge of meteorology, can sometimes provide clues to the source strengths. Had propane concentrations been seen to be similar in both hemispheres, for example, the data would have implied similar source strengths above and below the equator.

Probing Chemical Reactions in the Laboratory

This chapter speaks confidently about the occurrence of specific chemical reactions, and of some reactions progressing more rapidly than others. Readers who are not chemists might wonder how we really know these things. The answer is that the information comes from laboratory studies of reaction processes—a demanding and vital specialty in itself.

Consider, for example, the reaction of chlorine oxide (ClO·) with nitric oxide (NO),

$$ClO\cdot + NO \rightarrow Cl\cdot + NO_2 \qquad (3.20)$$

a process particularly important for stratospheric ozone chemistry. This reaction as written says that equal num-bers of extremely unstable ClO· radicals and relatively unstable NO molecules react to generate equal numbers of extremely unstable Cl· radicals and relatively unstable NO_2 molecules as products. Laboratory investigations of this reaction have several goals: to produce the reactants in high purity and at precisely determined concentrations, to allow the reactants to interact while monitoring the rate at which the interaction occurs and while minimizing any competing reactions that might confuse the analysis, and to detect the presence and measure the abundance of one or more of the reaction products to confirm that the reaction does indeed occur as written.

It is far from a trivial exercise to generate these highly reactive molecules and free radicals in a controlled manner. ClO· cannot be bought from a chemical supply shop, for example, so the experimenter must make it as a prelude to the reaction experiment itself. ClO· generation in the laboratory is a two-step process: chlorine molecules

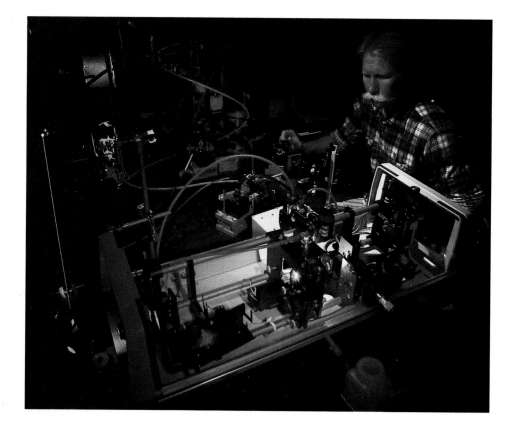

Dr. Craig Smith of the U.S. National Oceanic and Atmospheric Administration in Boulder, Colorado, uses laser light to detect short-lived chemical constituents in a laboratory study of atmospheric reaction rates.

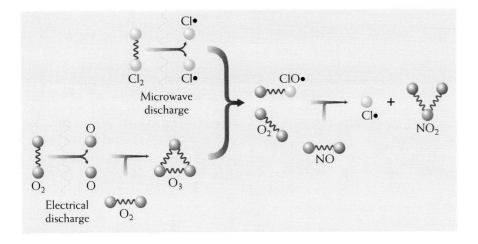

The sequence of steps involved in measuring the rate of reaction of chlorine monoxide with nitric oxide. The wavy lines indicate electrical discharges across the streams of flowing gases.

(Cl_2) are passed through a light source or microwave discharge, which breaks the molecular bonds and produces chlorine atom radicals ($Cl\cdot$); these must then be reacted with O_3, but you can't buy ozone either, so ozone is made by passing molecular oxygen (O_2) through a second light source or discharge, followed by reaction with a second oxygen molecule:

$$O_2 + \text{electrical discharge} \rightarrow O + O_2 \rightarrow O_3 \quad (3.21)$$

Then the two transient species are reacted to give $ClO\cdot$, one of the desired reactants, by

$$Cl\cdot + O_3 \rightarrow ClO\cdot + O_2 \quad (3.22)$$

These steps accomplished, NO can then be added (fortunately, it *can* be purchased, generally in pure enough form so that a purification step can be omitted), leading to the reaction that was the goal of the experiment in the first place:

$$ClO\cdot + NO \rightarrow Cl\cdot + NO_2 \quad (3.20)$$

In order to be certain that all this is happening as planned, and to measure the rate at which it is occurring, various detectors probe the system to measure concentrations of several of the molecules and free radicals involved.

These measurements, once analyzed, reveal the speed of the reaction and confirm the identity of the reaction products.

These analyses are extraordinarily arduous to perform, but nearly every important atmospheric reaction has been studied by several research groups around the world, often using different techniques, generally with less than 20% difference in the measured speeds of the reactions and with agreement on the identification of the reaction products. It is research such as this that supports our hypotheses and assures us that we understand the chemical reactions occurring in the atmosphere, high and low, near and far.

The Chemistry of Precipitation

"It's as pure as the driven snow," says a famous line from the classic musical *Guys and Dolls*. But how pure *is* snow? "Our shampoo is soft as rainwater," says the cosmetic advertisement. But how soft and pure is rainwater if it contains sulfuric acid? As we mentioned earlier, one of the ways certain trace gases are removed from the atmosphere is by absorption into rain and cloud droplets. What happens within those droplets when a gas is absorbed? If purity is the absence of contaminating substances, does such absorption mean that rain and snow are not so salutary after all? Have raindrops and snowflakes always been contaminated, without our knowing it?

Although some early chemical analyses of rain were carried out in England in the late 1800s, the first major modern study was conducted in the 1960s by Christian Junge, a German scientist on an extended appointment at the Air Force Cambridge Research Laboratory in Massachusetts, and later director of atmospheric chemistry at the Max Planck Institute for Chemistry in Mainz, Germany. Junge analyzed rain from a number of locations in North America, readily detecting the presence of chloride, sulfate, and nitrate ions in the water. His work, together with related research by University of Stockholm scientists, inspired numerous other studies with increasingly sophisticated instruments, and it has become obvious with time that the aqueous chemistry of the atmosphere—the chemistry involving droplets of water and flakes of ice and snow—is as complicated as the chemistry of the gas phase.

The single most important chemical species in clouds and precipitation is the hydrogen ion (H^+), whose concentration can be indicated by specifying the solution's acidity, or pH value. You may recall from high school chemistry that the pH scale ranges from 0 to 14, low pH values indicating high acidity (high concentrations of H^+) and high pH values indicating high alkalinity (low concentrations of H^+).

In pure water the pH is 7, and there are equal numbers of H^+ and OH^-, or hydroxide, ions. Water does not remain pure for long in the atmosphere, however, because the soluble gas carbon dioxide (CO_2) is present at a concentration of about 0.035%. It dissolves in water in a two-step sequence, first by incorporation

$$CO_{2(gas)} + H_2O \rightarrow CO_2 \cdot H_2O_{(dissolved)} \quad (3.23)$$

then by dissociation into H^+ and a bicarbonate ion (HCO_3^-):

$$CO_2 \cdot H_2O_{(dissolved)} \rightleftharpoons H^+ + HCO_3^- \quad (3.24)$$

The double arrow indicates that an equilibrium is established between the amounts of dissolved CO_2 molecules and their ions and that both phases exist in solution to some degree. Because the dissolution and ionization process has added H^+, the solution acidity has been increased. The gas phase concentration of CO_2 produces atmospheric droplets of about pH 5.6.

Most rain has a more acid pH than 5.6, largely because natural and anthropogenic nitrogen and (especially) sulfur species increase the acidity. The most abundant sulfur-containing gas to interact with rain is generally sulfur dioxide (SO_2), which dissolves and then reacts in solution to form HSO_3^-, the sulfite ion:

$$SO_2 + H_2O \rightleftharpoons H^+ + HSO_3^- \quad (3.25)$$

The sulfite ion is then oxidized by dissolved H_2O_2 or O_3:

$$HSO_3^- + H_2O_2 \rightarrow HSO_4^- + H_2O \quad (3.26)$$

$$HSO_3^- + O_3 \rightarrow HSO_4^- + O_2 \quad (3.27)$$

$$HSO_4^- \leftrightarrows H^+ + SO_4^{2-} \quad (3.28)$$

The nitrogen in acid rain comes not from nitrogen dioxide (NO_2), which is not very water-soluble, but from nitric acid, which is HNO_3. The latter is made from the former in the gas phase by reaction with our old acquaintance, the hydroxyl radical ($HO\cdot$):

$$NO_2 + HO\cdot + M \rightarrow HNO_3 + M \quad (3.29)$$

and then dissolved and ionized

$$HNO_{3\,(gas)} \xrightarrow{droplet} HNO_{3\,(dissolved)} \quad (3.30)$$

$$HNO_{3\,(dissolved)} \rightarrow H^+ + NO_3^- \quad (3.31)$$

Because of sulfur and nitrogen oxides, most rain near urban areas has pH levels nearer 4.0 than 5.6. Cloud and fog droplets are almost always even more acidic than rain. In some fogs, in fact, the pH of the droplets has been found to be as low as 1.7—close to that of battery acid! It is no wonder that vegetation and other materials exposed to such fogs deteriorate rapidly.

Acid rain became an international public policy issue at the first United Nations Conference on the Environment, held in Stockholm in 1972, where it was recognized that the SO_2 and NO_2 gases that are transformed into acids are often emitted in one locale while the rain falls in another, often across a state or national boundary. This situation was brought to light by the Swedes, who were downwind of airborne emissions originating in the British Isles and in other parts of Europe. Portions of the eastern

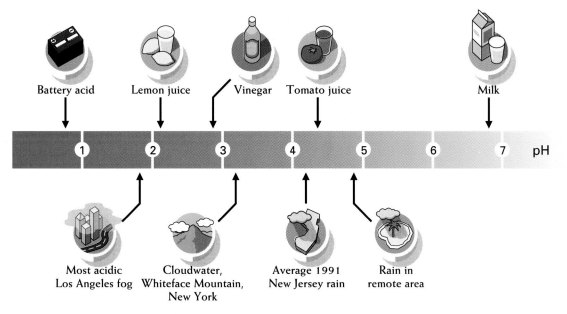

Battery acid Lemon juice Vinegar Tomato juice Milk

| 1 | 2 | 3 | 4 | 5 | 6 | 7 | pH |

Most acidic
Los Angeles fog

Cloudwater,
Whiteface Mountain,
New York

Average 1991
New Jersey rain

Rain in
remote area

The pH values in atmospheric water of various types, compared with the pH values for several common liquids.

Polyethylene funnels capture rain filtering through trees on Mt. Mitchell, North Carolina in an experiment studying the potential of acid rain to damage forests.

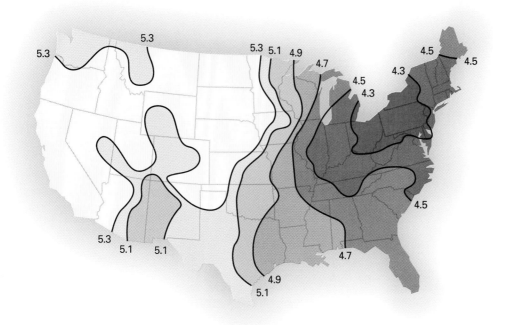

Lines of equal pH in rain for the continental United States in 1990. The levels below 5.0 east of the Mississippi River are the result of anthropogenic emissions of sulfur and nitrogen oxides.

United States and eastern Canada presented a similar scenario, the coal-burning power plants in the Ohio River Valley of the United States being the principal sources of emissions.

The discovery of acid rain was accompanied by the widespread assumption that the rain was causing substantial damage to soil, lakes, rivers, and structures. A decade or so of diligent research has confirmed some of these concerns and refuted others. The most severe and demonstrable damage appears to occur when highly acidic precipitation falls on surface waters within fragile ecosystems. The most sensitive of these systems are those of Scandinavia and northeastern North America, areas where the soil is thin and the organic matter that can absorb and neutralize the acids is sparse. In such waters and soils, sensitive organisms such as fish larvae are definitely affected. Certain tree species also show decline if they are simultaneously exposed to both acidic precipitation and toxic gaseous air pollutants such as ozone.

A second confirmed result of acidic precipitation is the irreversible decay of certain structural materials in buildings, automobiles, statuary, and the like, including many cultural resources. For works of art the damage is of particular concern because the objects are essentially irreplaceable. Vigorous conservation techniques, such as the periodic application of waxes to metal statues, can minimize this problem, but it is difficult to completely prevent degradation from occurring.

Other concerns about acid rain have been proved somewhat less well founded. There is little or no evidence that acid rain is affecting crops or human health, for example, and most lakes and forests seem adequately buffered by their rich organic ecosystems. Nonetheless, the damage that *is* occurring has been judged of sufficient magnitude to warrant the imposition of emissions controls on acid rain precursors, especially sulfur dioxide and NO_x. Such steps are currently being taken within Western Europe and North America. In the less developed countries of Asia, however, acidification problems are on the increase as a consequence of expanding rates of energy generation by the combustion of sulfur-rich coal, pollution control devices often being minimal or nonexistent. In those regions, cultural resources may be imperiled by local atmospheric chemistry for the next several decades.

Most studies of cloud precipitation acidity have been local or regional in scope, but atmospheric scientists are beginning to appreciate the need to study precipitation chemistry on a global scale, as well. For example, they are assessing the effects that clouds have on ozone production worldwide. Only about 10% of all clouds produce precipitation, the others evaporating as surrounding conditions change. Nonetheless, all clouds play important roles in atmospheric chemistry by virtue of their interactions with the air passing through them.

As soon as air enters a cloud, its chemistry changes in major ways. One of the reasons is a difference in the chemically reactive ultraviolet radiation that is received there. The main reason, however, is that critical gas phase constituents that are strongly water soluble, such as the free radicals $HO\cdot$ and $HO_2\cdot$ and the oxidizing molecule H_2O_2, become absorbed by the water droplets that make up the cloud. The less water-soluble components of the air mass, such as NO, CO, and CH_4, remain behind in the gas phase. This "filtering" process causes many of the reactions that have been taking place in the air—such as the ozone formation cycle, for example—to be strongly inhibited.

Jos Lelieveld and Paul Crutzen, both of the Max Planck Institute for Chemistry in Mainz, Germany, have estimated that air spends an average of 13 to 20 hours in a cloud-free environment followed by a 3- to 4-hour period inside of clouds. Taking account of both types of environments in a photochemical model, they calculated that the cloud-induced inhibition of gas-phase chemistry in the lower half of the troposphere leads to substantial reductions in ozone production there. For NO_x-rich regions, the reduction is about 40%. In NO_x-poor regions, net ozone destruction rates are enhanced by factors ranging from 1.7 to 3.7. As a consequence of lower ozone production, the concentrations of $HO\cdot$ are also substantially

The deterioration of marble due to acid rain is dramatically illustrated by photographs of this statue in Herten, Germany taken in 1908 (left) and 1969 (right).

A rainstorm washing pollutant particles and soluble trace gases from the air. Rain is a great cleaner of the atmosphere, once the hydroxyl radical and ozone have transformed poorly soluble pollutants into highly soluble reaction products.

reduced. An associated result is a significantly smaller rate of chemical removal of methane and other hydrocarbons from the air. Clearly, the atmospheric chemistry cannot be properly understood without a comprehensive knowledge of liquid-phase processes.

The Earth System's Sulfur Budget

In Chapter 2, budgets were presented for Earth's radiant energy and water cycles. Budgets are useful in studying individual elements or molecules as well. For example, scientists have used the budget approach to construct a picture of how sulfur is cycled through the Earth system. As with other budgets, to build a budget for sulfur it was necessary to know where it resides (the reservoirs), the processes that inject and extract sulfur from those reservoirs, and—to the extent possible—numerical values for reservoir contents and flows. For chemical budgets, such an analysis is often complicated by the need to look at nature's influences (such as volcanic emissions of sulfur) as well as those of humanity (such as emissions arising from the combustion of sulfur-containing coal).

Computing emissions to the environment is conceptually simple, as shown in the figure below. The emission sources are identified, the rate of emission from a typical source determined, and this rate multiplied by how often the activity occurs and how many similar sources are pres-

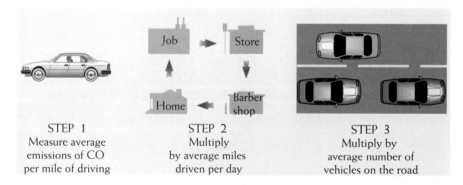

STEP 1
Measure average
emissions of CO
per mile of driving

STEP 2
Multiply
by average miles
driven per day

STEP 3
Multiply by
average number of
vehicles on the road

The process of computing emissions of carbon monoxide (CO) from automobiles in an urban area on a typical workday. The result will differ for each day of the week, season, age of vehicles, and other factors. For a complete inventory of CO emissions in the urban area, sources such as electrical power plants and industrial processes must also be taken into account.

Chapter Three

ent. However, the actual computation of emissions is as involved as the concept is simple. Fluxes generally vary by season, time of day, and weather, even when sources are operating as expected. As sources age or as they are fitted with emission control devices, their emission rates again require determination, and emissions computations must be adjusted appropriately. Computations designed to take all these variations into account are therefore quite complex. Emission factors must be separately determined for each emittant and each source, and because a substance may be emitted by multiple sources, the computations must sum for all that are relevant. The whole process is complicated enough to have become a subspecialty of atmospheric chemistry, and a high-quality regional or global inventory of a single chemical species can take a team of several people a year or two to complete.

Difficult as it is to make an accurate determination of anthropogenic emission fluxes, the evaluation of natural emission fluxes is more difficult still. Natural sources are generally uncontrolled and uncontrollable, their emissions complicated to measure, and the source locations highly variable and often geographically remote. For some chemical species, such as the hydrocarbons emitted from different groups of trees, emissions may vary as a function of temperature, soil nutrients, and soil moisture—factors that may be imperfectly known.

A rational reaction to this discussion of emissions estimation is to wonder if it can be done. The answer is a guarded yes, with the proviso that one must be willing to accept that emissions inventories always have built-in uncertainties, sometimes quite high ones. The many inventories that are now constructed with the cooperation of

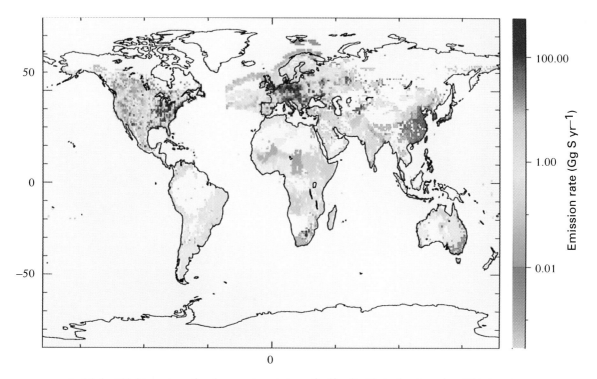

A global gridded inventory of anthropogenic emissions of sulfur dioxide to the atmosphere. This inventory was prepared for reference year 1985 and includes industrial and combustion-related releases.

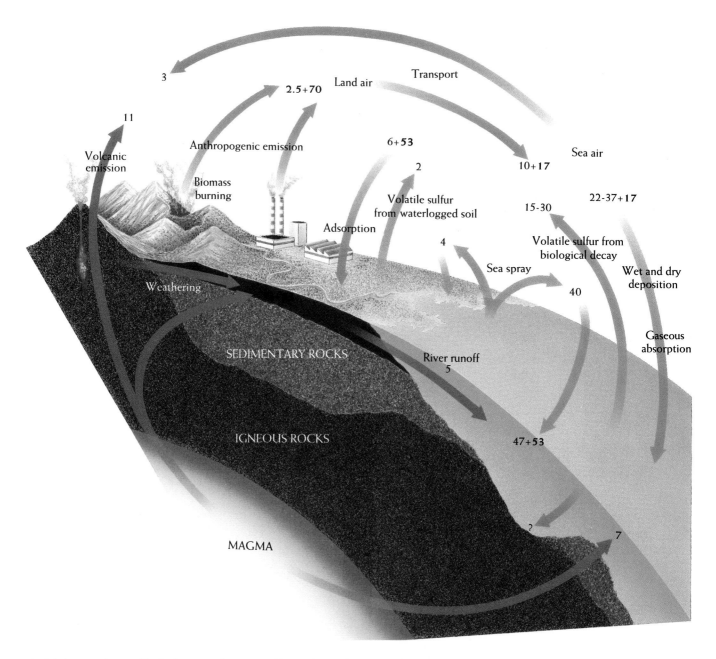

Volcanic
emission

11

Anthropogenic emission

2.5+**70**

Biomass
burning

Land air

Transport

3

6+**53**

2

Volatile sulfur
from waterlogged soil

Sea air

10+**17**

22-37+**17**

15-30

Adsorption

Volatile sulfur from
biological decay

Wet and dry
deposition

Weathering

4

Sea spray

40

Gaseous
absorption

SEDIMENTARY ROCKS

River runoff

5

IGNEOUS ROCKS

47+**53**

MAGMA

2

7

A global atmospheric sulfur budget. Numbers in small type denote the estimated natural contribution; numbers in larger bold type denote the anthropogenic contribution. The units of the fluxes are teragrams (million metric tons) of sulfur per year.

large numbers of scientists are considered to be works in progress, continually improving but never complete. Nonetheless, the task of constructing them is crucial, not only for scientific purposes but because any regulatory actions aimed at controlling emissions must be founded on a realistic idea of the level of emissions and of their natural and anthropogenic sources: we need to know what it is we may have to control.

Once scientists have constructed an emissions inventory, assessed transport and transformations, and measured or estimated loss rates, they are ready to construct a budget and see if it is in balance, that is, to determine whether an element or compound of interest is being gradually transferred from one reservoir to another, or whether its flows from the reservoirs are compensated by flows into them. An example assessment for a global sulfur budget is illustrated on the facing page. Experts generally concede that the sulfur budget is dominated by industrial activities, which result in the emission of some 70 teragrams of sulfur per year (1 Tg = 1 million metric tons) to the atmosphere, almost entirely as sulfur dioxide. Of this amount, about 25% is thought to be deposited over the oceans, while the remainder is deposited on land. Natural biological emissions of sulfur on the continents amount to only about 2 Tg of sulfur per year, but the emissions of sulfur from the oceans are much larger. There it is produced by microorganisms, which emit it in the gas dimethyl sulfide (CH_3SCH_3) at the rate of 15 to 30 Tg of sulfur per year. (Note the considerable uncertainty about this rate, as is often the case for natural emissions.) Most of the oceanic sulfur is returned to the oceans after being converted to sulfuric acid and dissolved in precipitation, but perhaps 10% is transported to the continents by surface winds. Other natural sources of atmospheric sulfur are volcanoes and fumaroles. These sources are quite erratic, and the estimated average annual flux of 11 Tg of sulfur is very uncertain. Because most of Earth's volcanic activity occurs on islands or at oceanic margins, much of the sulfur from the volcanic eruptions is subsequently deposited over the oceans.

What do we make of the global sulfur budget now that we have it in hand? One message is that some parts of it appear rather well determined, others not. The budget thus serves as impetus for scientists to make better measurements of such factors as volatile sulfur from biological decay and sulfur in precipitation ("wet deposition"). A second message is that the results may be used to provide rough estimates of the effects of acid rain on soil and statues, since a budget provides an idea of how much sulfur is transferred from air to rain and hence to the ground. Finally, the budget—if good enough—can address the crucial questions: Are inputs and outputs in balance? Even if they are, are they correct? (A balanced budget may have errors in inputs and outputs that compensate for, and thus disguise, each other, or an unknown term may wrongly be supposed to be large enough to provide budget balance.) Are reservoir contents rising or falling? Which factors are the critical ones in achieving a balanced atmospheric budget? Such questions are central to atmospheric research, and we will return to them in subsequent chapters.

Sedimentary outcrops in Spitsbergen, Norway. Exposed rock layers such as these, deposited hundreds of millions of years ago, provide information concerning ancient climate and the evolution of life on Earth.

Climates of the Past | 4

To a person uninstructed in natural history, his country or seaside stroll is a walk through a gallery filled with wonderful works of art, nine-tenths of which have their faces turned to a wall.
—Thomas Henry Huxley

How did the atmosphere, the creator and mediator of climate, come to be, and how has it evolved over the eons? This question, asked by innumerable scientists and laypersons alike over the centuries, has instigated one of the great true detective stories of history, complete with all the elements of the classic English mystery novel. For example, in the novel, after a murder victim has been discovered, the detective proceeds to examine all available evidence—the victim's clothing, signs of violence, information from family members, and so forth, to reconstruct the patterns and distinctive features of the victim's life, especially at certain crucial moments in time. In much the same way, paleogeologists and paleoclimatologists gather rocks of all ages, trapped air bubbles, water frozen in ice sheets, fossils from the distant past, and even the testimony of relatives (the other planets). From this diverse and often confusing collection of information, they attempt to reconstruct a picture of the birth, young adulthood, and maturity of planet Earth.

A key discovery of the paleogeological detectives is that nowhere is the concept of an Earth system—intimately connecting the processes of biology, geology, hydrology, and atmospheric science—better illustrated than in the story of the formation and evolution of Earth's atmosphere, because each of those four realms has contributed to the result. The sequence began with the transition of the Sun from an embryonic and then juvenile star to a mature one, a process that was accompanied by the gravitational accretion of gas, grains, and dust in the solar nebula to form the planets of the solar system. Various smaller solid bodies—moons, asteroids, comets—were formed as well, often with highly perturbable orbits. Theoretical analyses suggest that the collapse of the solar nebula and the accretion of dust and gas to form Earth and other bodies of the solar system occurred within the geologically short time span of perhaps 50 to 100 million years.

The Birth of Planet Earth and Its Atmosphere

In the last two decades, spaceborne and ground-based instruments have allowed us to compare many of the characteristics of the neighboring planets Earth, Venus, and Mars. At present, these planets have very different chemistries and climates. Compared with Earth's atmosphere, that of Venus has less water, much less molecular oxygen and molecular nitrogen, a very high concentration of carbon dioxide, and a very high relative concentration of sulfur dioxide. The Martian atmosphere is also dominated by carbon dioxide; it has little molecular oxygen and molecular nitrogen, very little water, and no detectable sulfur. The surface temperatures of the three are dramatically different as well, Venus being hundreds of degrees warmer than Earth, and Mars about seventy degrees colder.

Are the differences in chemical composition a result solely of the relative positions of these planets with respect to the Sun? Many experts believe that the differences arise in part from the relative retention of volatile substances (water molecules and carbon, chlorine, sulfur, and nitrogen atoms) by the solid particle grains that accreted to form the planets, and this quality is indeed related to the distance of the planet from the Sun. For example, a planet forming near the Sun is less likely, because

of its higher temperature, to have minerals such as talc or serpentine, $Mg_3Si_4O_{10}(OH)_2$ and $Mg_3Si_2O_5(OH)_4$, respectively, which have water incorporated into their crystal structures. Thus, the minerals of Venus probably formed with little water trapped in their interstices, Earth with more, and Mars with more still. Temperature and mineral considerations suggest that Venus and Earth originally contained similar amounts of carbon dioxide, chlorine, and sulfur, whereas Mars had more sulfur and chlorine than Earth but less carbon dioxide.

Once the planets were formed, the naturally buoyant gases in the planetary interiors were expelled by volcanism to form the initial atmospheres. Having a higher temperature, Venus probably expelled more of the gases that were contained in its interior than did Earth. Because little water was present on Venus, much of this outgassing consisted of carbon dioxide and sulfur dioxide, and the large amounts of these gases in the atmosphere immediately absorbed infrared radiation and set up a "runaway" planetary greenhouse. Without water to dissolve the carbon dioxide, or plants and animals to recycle it, the greenhouse on Venus stabilized at an extremely high temperature, much too warm for life as we know it.

The opposite extreme is exemplified by Mars. Although some outgassing must have occurred on that similar, cooler planet, carbon dioxide and water in the Martian atmosphere were apparently never abundant enough to produce much of a greenhouse effect, so the planet never became warm enough to support life. Furthermore, without significant liquid water, many of the dissolution and precipitation processes that make Earth's climate so hospitable to life were never able to develop.

Thus, Earth appears to have been especially favored from birth as a potential reservoir for life. Its early atmosphere was rich in hydrogen, a remnant of the original solar gases and dust that coalesced to form the solar system. Even so, many other members of the solar system also have high abundances of hydrogen but do not possess the abundance of molecular oxygen so important to life on Earth. How did our oxygen-rich atmosphere come about?

Most of the evidence scientists have relied on for clues to the composition of Earth's early atmosphere is necessarily circumstantial, but we can at least paint a plausible scenario. The planet had reasonable supplies of water

Physical Characteristics
of the Terrestrial Planets

Characteristic	Venus	Earth	Mars
Total mass (10^{27} g)	5	6	0.6
Radius (km)	6049	6371	3390
Atmospheric mass (ratio)	100	1	0.06
Distance from Sun (10^6 km)	108	150	228
Solar constant (W m^{-2})*	2613	1367	589
Albedo (%)	75	30	15
Cloud cover (%)	100	50	Variable
Effective radiative (°C) temperature	-39	-18	-56
Surface temperature (°C)	427	15	-53
Greenhouse warming (°C)	466	33	3
N_2 (%)	< 2	78	< 2.5
O_2 (%)	<1 ppmv	21	< 0.25
CO_2 (%)	98	0.035	> 96
H_2O (range %)	$1 \times 10^{-4} - 0.3$	$3 \times 10^{-4} - 4$	< 0.001
SO_2 (fraction)	150 ppmv	< 1 ppbv	Nil
Cloud composition	H_2SO_4	H_2O	Dust, H_2O, CO_2

*The solar constant is the intensity of the solar radiation over a square meter of surface at a distance equal to that from the Sun to the planet's orbit.

locked within its forming particles, and Earth was located at an appropriate distance from the Sun for that water to exist in liquid form. Even given its supply of water, though, a serendipitous amount—enough but not too much—of infrared-absorbing gases was apparently necessary for maintaining surface temperatures within the liquid water range. Thus, carbon dioxide (CO_2) is thought to have had an important role in establishing the principal features of Earth's climate. The other major atmospheric constituents were nitrogen (N_2), hydrogen (H_2), and water (H_2O), with perhaps smaller amounts of hydrogen sulfide (H_2S), ammonia (NH_3), and methane (CH_4). There was origi-

nally no free oxygen. Because iron oxide (FeO) was readily available, it is hypothesized that additional H_2O formed in a process represented as

$$FeO + H_2 \rightarrow Fe + H_2O \qquad (4.1)$$

with the H_2O remaining at the surface or in the atmosphere and the free iron (Fe) gradually collecting in the planet's hot core and mantle.

Early in the planet's life cycle, and within an atmosphere like that described above, the first forms of life appeared. These life forms were bacteria that lived by break-

Precambrian "red bed" deposits of limonite (hydrated iron oxide, the rust-colored layers) at Hammerslee, Australia. It is thought that red bed formation early in Earth's history extracted oxygen from the atmosphere, thus delaying its accumulation there.

ing down molecules under oxygen-free conditions. Eventually, about 3.8 to 3.5 Gyr BP, some of the bacteria began to develop a photosynthetic capability (the ability to synthesize chemical compounds by using solar energy to drive the reactions) that made use of H_2O and CO_2 instead of molecules not containing oxygen. The hydrogen in the H_2O and the carbon in the CO_2 were used to form molecules useful to the bacteria themselves, such as formaldehyde (HCHO, the building block of sugar); the O_2 was a waste product, ultimately escaping to the atmosphere. The process may be represented as

$$CO_2 + H_2O \rightarrow HCHO + O_2 \qquad (4.2)$$

Molecular oxygen was also being slowly produced by the photodissociation of water in the atmosphere, followed by the loss of the light hydrogen atoms to space at altitudes near 500 km. For a long period, any oxygen released near the ground was immediately captured by iron ions dissolved in the water, a process reflected in the ancient "red bed" deposits of oxidized iron found in northern On-

tario, Canada, the petrified forests of Arizona, and many other locations across the globe. Eventually, however, the rate of oxygen production exceeded the rate at which iron was brought to Earth's surface by geological processes. At that point, some 2 gigayears ago, the oxygen began to accumulate in the atmosphere. This change brought about a biological revolution: the oxygen-rich atmosphere was lethal to most existing biological species, but it stimulated the development of other species that were able to utilize O_2, an approach that also uses solar energy much more efficiently. Paleoatmospheric chemists have dated three significant milestones in the evolutionary history of molecular oxygen:

2.0 Gyr BP: atmospheric O_2 reaches 1% of present value

700 Myr BP: atmospheric O_2 reaches 10% of present value

350 Myr BP: atmospheric O_2 reaches 100% of present value

The accumulation of O_2 gradually produced O_3 by the absorption of solar radiation, utilizing the same chemical reaction sequence operating in today's atmosphere:

$$O_2 + h\nu \rightarrow O + O \qquad (4.3)$$

followed by

$$O_2 + O + M \rightarrow O_3 + M \qquad (4.4)$$

The increasing levels of atmospheric O_2 and O_3 began to shield Earth's surface from the lethal solar ultraviolet radiation, eventually permitting biological evolution to occur on land as well as in the sea. With the existence of the water-rich, oxygenated atmosphere and surface, Earth's climate and atmospheric chemistry assumed major roles in the planet's physical, chemical, and biological development and were themselves, in turn, affected as the changes they helped bring about reached global scales.

So goes the scenario. But what evidence is it based on, and how much of it can we prove as we try to deduce the atmospheric conditions of the distant past?

Approaches to Ancient Climates

The eras of geologic time, segregated from each other by great events in Earth system history, are distinct as well in their atmospheric and climatic characteristics. The earliest era, the Precambrian, ended with the first great burst of life at about 570 Myr BP. Its successor, the Paleozoic, is defined as terminating at 225 Myr BP, at which time today's continents began to form and drift. The subsequent era, the Mesozoic, extended to 65 Myr BP, the time of the sudden extinction of many of Earth's life forms. For pedagogical convenience, we divide the modern era, the Cenozoic, into two parts: the pre-Holocene epoch (65 Myr to 10 kiloyears [kyr; 1 kyr = 10^3 years] BP, a time period

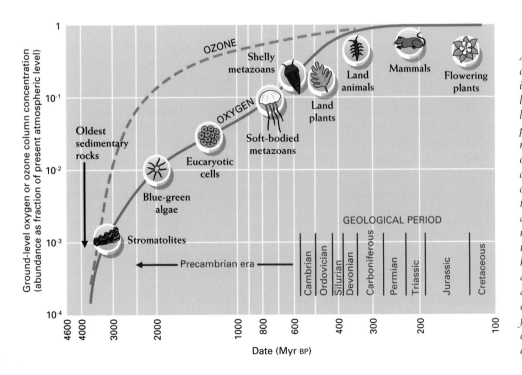

A reconstruction of the evolutionary development of oxygen and ozone in Earth's atmosphere. Land-based life could not have become established without sufficient ozone to provide protection against ultraviolet radiation. The times of appearance of several key life forms are indicated: stromatolites are a type of carbonate rock believed to be indicative of primitive algae, blue-green algae are undeniably similar to modern algae, eucaryotic cells are those which have nuclei (as do highly evolved bacteria and more advanced life forms), and metazoans are multicellular organisms. The oxygen curve is based on evidence from rocks and fossils; the ozone curve was developed from a photochemical computer model.

for which we know a modest amount about climate and chemistry), and the Holocene epoch (10 kyr BP to the present), a time period about which we know a lot.

Paleoclimatology, the science that studies the climates of the distant past, consists basically of three separate activities. The first is the search for objects that were formed at about the time one wishes to study and that reflect the climate at the time of their creation. The second is to date the objects accurately. The third is to draw inferences about the climate on the basis of information extracted from the objects. The further back in time we go, of course, the less information (the fewer objects) we have to draw on. For the Precambrian interval (4.6 to 0.6 Gyr BP), the stretch of time from Earth's initial formation to the first appearance of marine animals with skeletons, the sole record consists of rocks that have been preserved from erosion and rescued from burial. It is these rocks that we study to learn about the climate early in the history of our planet.

Although Earth system scientists speak glibly these days about various stages of Earth history, the ages of old rocks and of the planet itself have been known with reasonable accuracy for less than a hundred years. The key to determining those ages was the discovery of radioactivity in Earth rocks by French physicist Henri Becquerel and German physicist Wilhelm Roentgen in the last decade of the nineteenth century.

The technique of radioactive dating consists of measuring and comparing the relative concentrations within a sample of parent nuclei (the radioactive species that undergoes disintegration—uranium is one example) and daughter nuclei (the species that are produced by the disintegration—lead is a daughter element to which uranium decays). Because the rate of decay for the parent species is known, one can calculate the time when only parent atoms were present, that is, the time when the rock originally formed. Once this decay process was sufficiently understood, it was realized that paleogeologists could use it as a clock; as such, it is the key that opens the door to the climate history of the young Earth.

Although a number of radioactive parent-daughter combinations can be used for determinations of age, most of the epochs identified in Earth history have been dated through the use of the uranium-lead system, in which the decay is sufficiently slow that it opens to us the earliest time periods. The analytical techniques used to detect the parent and daughter nuclei are so sensitive that they can be applied to grains small enough to be visible only under a microscope; it is thus possible to date extremely tiny fragments remaining from the young Earth. Another application involves measurement of radioactive carbon, continuously formed in small amounts in the atmosphere by cosmic rays and incorporated into leaves and grass in photosynthesis just as is normal carbon. Through the combination of uranium and carbon dating with that from potassium, thorium, and other naturally radioactive elements, paleogeologists can provide dates for rocks and fossils representing nearly the entire sweep of Earth's history.

Precambrian Climate

When radiometric techniques are applied to grains in rocks thought to be extremely ancient, the oldest are dated at about 4.2 Gyr BP. These grains are chemically typical of ocean sediments and are rounded in the classical physical pattern of sedimentary deposits, indicating that the grains were laid down under water. The presence of water indicates that average temperatures exceeded 0°C at the time and in the vicinity where the sediments were deposited, and the presence of such grains widely dispersed over the planet indicates that surface waters were common throughout almost all of Earth history. The presence of surface waters also implies the existence of at least a modest atmosphere, because liquid water evaporates rapidly at low atmospheric pressure. As a consequence, except for the first 400 million years or so of the planet's life, for which we are likely never to have direct evidence, we can say with some certainty that the planet has always had substantial amounts of water coverage and at least a modest atmosphere. Furthermore, carbonate minerals—those that contain a carbonate (CO_3) constituent, as does limestone, which is mostly $CaCO_3$—are seen in rocks of all ages. These minerals precipitate from oceans that contain dissolved carbon dioxide in equilibrium with an atmosphere that contains gaseous carbon dioxide; we therefore deduce that carbon dioxide has been a constituent of the atmosphere since the very early stages of the planet's existence.

The lower limit for Earth's surface temperature during the Precambrian era is well-enough defined merely by water's presence over much of the globe, but not so its upper limit. It is possible, however, to derive an upper limit to the temperature from the presence of the oldest known single-celled fossils, in rocks from about 3.8 Gyr BP, and of more extensive biological communities at least as early as 3.5 Gyr BP, whose growth temperatures must have been near those of similar organisms in the present world. Also of use as paleothermometers are temperatures derived from ancient tillites, rocks deposited by glaciers as they advance and retreat and which are formed only under rather rigid temperature constraints. As illustrated in the graph below, these considerations suggest that our early water-covered world would have had reasonably stable temperatures averaging perhaps 7°C, although temperatures a few degrees below the freezing point of water in parts of the planet are not ruled out.

As Earth's distance from and inclination toward the Sun undergo their periodic changes (more on these later), as the oceans ebb and rise, and as continents are formed, the albedo, the reflectivity of the Earth's surface, also undergoes change. Since radiation that is absorbed warms Earth and radiation that is reflected cannot, such changes are reflected in climate. Land areas influence the albedo to greater or lesser degrees through events such as ice cap formation (making the surface more reflective) and growth of vegetation (making it less reflective). The records of periods in which large areas of land have been covered by snow and ice are retained in the sizes, types, conditions, and positions of rocks. This evidence has been used to show that Earth has passed through a number of cycles of glaciation (the formation and subsequent removal of substantial ice cover).

One particularly active area of experimental geology is the study of temperatures, pressures, and physical and chemical environments under which different minerals form. In the most dramatic examples of that pursuit, minerals found at Earth's surface are compressed in laboratory anvils to pressures a million times that of today's atmosphere at sea level and heated by lasers to a thousand times the surface temperature of Earth, conditions that mimic those at the boundary between Earth's molten iron outer core and its silicate mantle. Only somewhat less demanding are related studies of the grinding action and temperature and pressure conditions peculiar to glaciation. When rocks demonstrating such chemistry and structure are radioactively dated, the results provide a history of ancient glaciation, limited only by the degree of completeness of the rock record. The earliest verifiable glacial epoch (but not necessarily the earliest glacial epoch that occurred) is at about 2.7 to 2.3 Gyr BP. The glaciation appears to have been extensive, although the supporting record is quite fragmentary. Its cause is uncertain; it may have been a consequence of rather low solar luminosity (a phenomenon we discuss later), of the presence of significant ice-coated

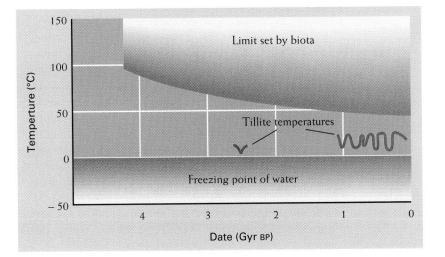

Constraints on the history of temperature near Earth's surface. The strongest constraint is imposed by the continuous existence of liquid water, which requires temperatures exceeding 0°C at least locally. The next strongest constraint is probably the occurrence of certain distinctive rock features (such as tillites at 2.2 Gyr BP and more recently, indicated by the squiggly lines). The biota provide an upper constraint in that large groups of organisms are known to be intolerant of high temperatures; those tolerances decrease with increasing evolutionary rank, and a widespread and diverse biota of a given rank implies a fairly rigid constraint on ambient near-surface temperature.

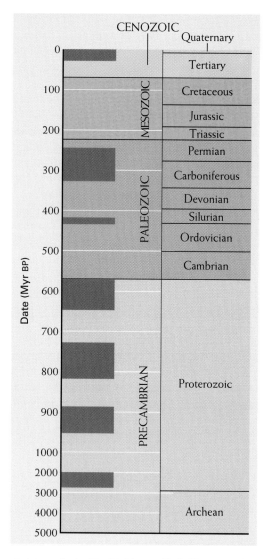

Major glacial epochs in Earth's history (dark blue regions). Note that the time scale changes at 1000 Myr BP.

and two others followed at about 820 to 730 Myr BP and 640 to 580 Myr BP. The late Precambrian was a major period of mountain-building on Earth, and the glaciations may have been related to continental motions (continents moving toward the poles are more likely to be ice covered) and topographic disruption (glaciers form preferentially at high altitudes).

A mysterious feature in Earth's early climate is the "weak Sun paradox." The Sun is a common type of star whose life cycle and properties are thought to be well known. Except for a brief period very early in the formation of such a star, the standard models of solar physics indicate that the luminosity (the rate of total energy output) of the early Sun was about 70 to 80% of the present value and that it has increased steadily since that time. As a result, the amount of solar radiation received by the young Earth would have been much less than is being received today, and the effective radiation temperature of the planet would be expected to be 10 to 15°C below today's value. Nonetheless, evidence from the rock record indicates that the surface temperature of the Precambrian Earth was quite similar to that of today.

How could significantly less radiation have produced a warm Earth? One hypothesis posits a greater retention of infrared radiation, and the most plausible way in which that could happen would be if a higher percentage concentration of carbon dioxide were present in the atmosphere at that time than now, mainly because the embryonic biosphere of that period was not utilizing as much carbon dioxide and producing the same amount of carbonate sediments as the biosphere of today. A second suggestion, inspired by speculative new models of solar physics, is that the early Sun (and other stars of its type) may have lost mass as it evolved and that this loss resulted in a high early solar luminosity. The higher luminosity calculated in this mass-loss model could easily explain the existence of early planetary temperatures in the liquid water range, but the model's reliability is not yet clear.

The resolution of the weak Sun paradox—how a weaker Sun produced an unexpectedly warm planet—is not likely to occur soon, and we may never know for certain the early workings of the Earth system. What does appear certain, however, is that although we cannot at present explain why, most or all of Earth was covered with liquid water during its juvenile and young adult life.

land masses to reflect radiation, and of low concentrations of greenhouse gases.

Following this early glaciation (at about the junction of the Archean and Proterozoic periods), the absence of glacial rock for the next billion or so years indicates that Earth was warm and devoid of permanent snow or ice. The second known glaciation occurred at about 950 Myr BP,

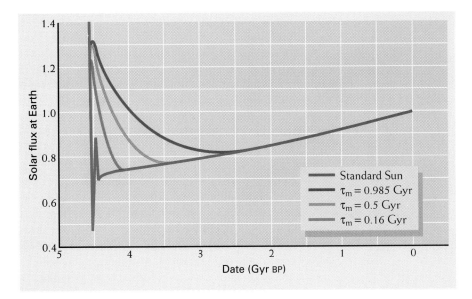

The solar flux (energy per unit area and time) at the distance of Earth (relative to the present solar radiation) resulting from a standard Sun model and from three models all beginning with a mass 1.1 times that of the present Sun and allowing mass loss. Three different time scales of mass loss (τ_m) are shown here. At present, there are few data to indicate which of these might be the most accurate.

Paleozoic and Mesozoic Climate

The erosive action of rivers and the thrusting of colliding continents have exposed layered rock in many areas throughout the world. Stratigraphy, the study of the individual layers (strata) and their constituents, provides at least an approximate historical record of climate and life, since the layers formed sequentially as particles carried by air or water were deposited on Earth's surface and then became consolidated into sediments that hardened over time into rock. Contained within this rock are the chemical and fossil records revealing the conditions under which the sediments were created.

The stratigraphic fossil record, established in the late eighteenth century as the relative order of the appearance of organisms through time, was correlated with radiometric rock dates by British geologist Arthur Holmes in the 1910s, thereby linking reliable dates with an extensive independent body of evidence about ancient life and climate. One of the longest-recognized and most distinctive features of the fossil record has been dated at about 570 Myr BP: the layer of rock in which marine animals with shells make a sudden, explosive appearance. The era of the subsequent rapid evolution and proliferation of these animals, the Paleozoic, is represented by relatively abundant rocks found throughout the world.

Ancient climate can be determined from the fossil record because fossil shells are made of calcium carbonate ($CaCO_3$), a compound containing oxygen atoms that were originally incorporated from the sea water the animals lived in. Today, those oxygen atoms provide clues to the temperature of the water at the time the shells were formed. The two common isotopes of oxygen are ^{18}O and ^{16}O, the latter being lighter by two neutrons. Both liquid water and dissolved carbon dioxide contain oxygen, and thus the two oxygen isotopes, and the isotopes distribute themselves between H_2O and CO_2 so that the energy of the system is at a minimum. Fortuitously for paleoclimatologists, the distribution of the isotopes is a function of the temperature of the water. Analyses of the oxygen isotope ratios in carbonate fossil shells thus provide a signature of the climate in which each shell was formed. Comparisons of these measurements from Paleozoic rocks around the world reveal the geographical extent of various contemporaneous climates.

One major climatic event of the early Paleozoic era revealed by the rocks and fossils is a brief glaciation that occurred at about 430 Myr BP. This event immediately preceded the evolution of land plants at the relatively re-

Layers of rock formed by the solidification of sediment and then folded by the motions of Earth's crust. Analyses of the rocks reveal information about Earth's climate and chemistry at the time of the rocks' formation.

which for all practical purposes continued to confine them to moist habitats. By 350 Myr BP, however, plants had developed their second great evolutionary innovation: seeds. Seeds, fertilized within the plant and small enough to be carried by wind or water, liberated plants to explore and colonize dry land.

The widespread growth of plants produces vegetative land cover that, when fully developed, can decrease the albedo by as much as 10 to 15%. Once the albedo is decreased, the heating of the land by sunlight is much enhanced, and the entire climate is revolutionized. Perhaps this is why the Paleozoic era was generally warm before concluding with a long and somewhat mild glacial period, centered at about 300 Myr BP. This glaciation was probably related to continental drift, which carried some of the largest continents along paths that traversed the poles.

Rather abundant climatic information is available for the time period beginning about 170 Myr BP, when sediment that still remains largely undisturbed was laid down on the bottoms of the oceans. Many samples of this material have been obtained as part of the Deep Sea Drilling Program, an international scientific cooperative effort that is recovering cores of rock from holes drilled deep

A diorama reconstructing the appearance of primitive land plants during the Devonian period, 395–345 Myr BP.

cent date of about 400 Myr BP. The first land plants resembled rigid stems, probably lived in marshes, and were closely related to waterborne vegetation. Two major evolutionary developments allowed them to migrate out of the water. The first was the plant vascular system, in which one set of tubes carries water and nutrients upward from the roots while another set transports and distributes the food manufactured in the leaves from atmospheric carbon dioxide. The earliest vascular land plants roughly resembled the primitive ferns of today; they reproduced by releasing spores into the environment, a mechanism requiring the released sperm to encounter the released egg,

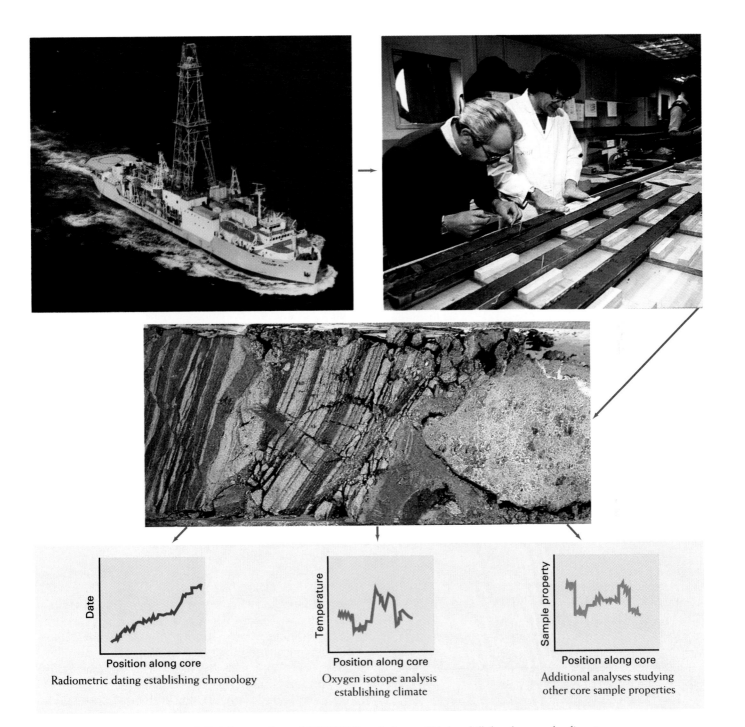

(Top left) *The research vessel* JOIDES Resolution, *which has drilled and recovered sediment cores from more than 300 locations around the world.* (Top right) *Scientists examine a portion of a sediment core drilled from the ocean floor.* (Center) *Closeup of a typical sediment core section.* (Bottom) *Schematic diagrams of some of the information produced by the scientific analyses of marine sediment cores.*

Radiometric dating establishing chronology

Oxygen isotope analysis establishing climate

Additional analyses studying other core sample properties

69

into the ocean floor. Unlike exposed rock layers on land, whose records are invariably complicated by such factors as folding, faulting, and incompleteness, marine cores permit a detailed examination of the climate of the Mesozoic era, which spanned the period of 225 to 65 Myr BP and appears to have been another long interval of widespread warmth, especially in the higher latitudes. Globally, temperatures were some 6 to 10°C warmer than at present. Throughout much of the Mesozoic era, the continents were joined in a single supercontinent that extended from high southern to high northern latitudes. This geographical situation may have facilitated the transport of heat by ocean currents from the equator toward the poles, thus resulting in a more even global temperature distribution than would have been the case with dispersed land masses.

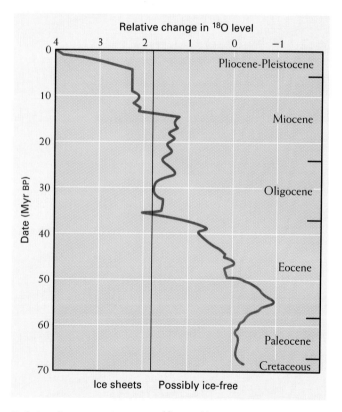

Relative changes in the ratio of ^{18}O to ^{16}O concentrations in marine fossils from Cenozoic era samples recovered by the Deep Sea Drilling Program.

As the continents formed and collided and large land masses began to grow, substantial gradations in altitude were created on Earth's surface, polar ice caps formed, and the ratio of water to land decreased. The Indo-Australian plate's collision with the Asian plate at about 50 Myr BP resulted in the formation of the Tibetan plateau, the most dramatic topographic feature of today's world. Roughly coincident with this collision, the climate underwent rapid cooling and large ice sheets eventually formed. The Tibetan plateau is so high and wide that it influences the atmospheric circulation of the entire northern hemisphere. The changes in precipitation and temperature that resulted from its creation are seen in a variety of ancient geological records, such as the oxygen isotope content of fossils in the deep sea.

Much of the history of water coverage on the planet is derived from studies of the present continental structures, because the evidence for sea level change is most accessible on their margins. These studies exploit the fact that sequences of sedimentary rock formations that develop inland are different from those that develop at the ocean margins, so that by determining the ages and locations of each type of sequence, one can tell whether a given location near a continental margin was or was not underwater at a particular time. If similar results are obtained for samples taken from different locations over a large area, then regional sea-level changes can be separated from purely local ones; if similar results are obtained for a number of regions over the entire planet, then the underlying cause can be considered to be eustatic, that is, associated with a worldwide change in sea level.

Eustatic curves constructed by the methods outlined above trace the history of sea level from Cambrian time to that of the present, as shown on page 73. The data show that throughout much of recorded Earth history the seas have been far higher than they are at present. In addition to causing widespread flooding of the edges of continents, such conditions would have produced considerable flooding of continental interiors, creating the vast inland lakes whose existence has been well documented by paleogeologists.

What is the cause of eustatic changes? Two reasonable possibilities have been suggested: one is the melting and freezing of polar caps; the other is changes in the volume

A

B

C

D

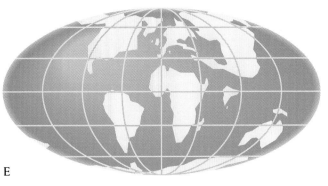

E

F

Reconstructed locations for the continents. (A) 320 Myr BP: most of the land was contained in two supercontinents, Laurussia (combining Greenland, North America, Scandinavia, and most of Russia) and Gondwana (containing most of the rest of the land mass). (B) 250 Myr BP: the collision of the two supercontinents formed Pangaea, a new supercontinent; the mountain ranges created in the suture zone included the Appalachians. (C) 135 Myr BP: Pangaea begins to break up, and the Atlantic Ocean starts to form between North America and Eurasia. (D) 100 Myr BP: Africa, South America, and India separate. (E) 45 Myr BP: Africa, South America, and India move northward toward Europe, North America, and Asia, respectively. Australia breaks off from the southern land mass. (F) Today: the American continents are joined, as are Africa and Eurasia. India has collided with Eurasia to form the Himalayan mountains.

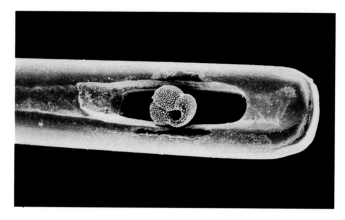

Foraminifera, radiolaria and sponge spicules extracted from a Deep Sea Drilling Project core.

A sediment core (left) from the North Atlantic reflects the abrupt change in ocean circulation that occurred some 130,000 years ago. A scanning electron micrograph of the lower, darker portion of the core (bottom center) reveals rock fragments rich in silicon (blue in the X-ray map at (bottom right) dropped by melting icebergs. The upper, lighter portion of the core is comprised largely of shells (top center) from marine animals inhabiting warmer waters. The shells are rich in calcium (red on the X-ray map at top right).

Chapter Four

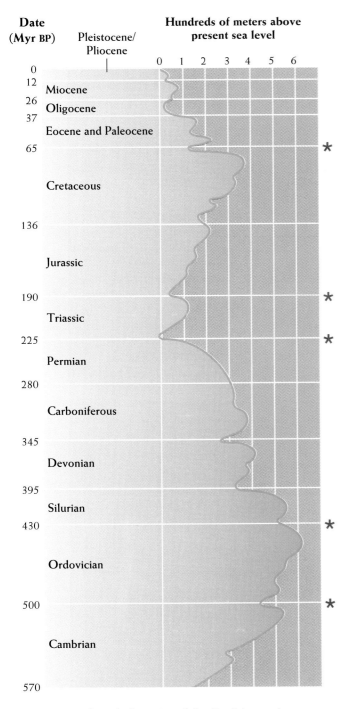

Date (Myr BP)	Pleistocene/Pliocene	Hundreds of meters above present sea level

Eustatic curves from the beginning of the Cambrian to the present day. Great marine biological extinction events are indicated by asterisks.

of ocean basins. Records of the polar ice caps are contained in the rock and debris that indicate the extent of glaciers; those records show that glacial expansion and contraction is insufficient to explain the eustatic changes seen in the data. Changes in ocean basin volume, therefore, turn out to be a likely cause. As we outlined earlier, tectonic activity results in enhanced seafloor spreading, the formation of substantial midocean ridges, and hence the interaction of plate tectonics and eustasy. The tectono-eustatic processes responsible for the sea level history result from variations in heat flow from the planetary interior to Earth's surface, and the variations in eustasy can be explained in at least approximate terms by the interaction of internal heat with stress and strain relief in the overlying rock structures.

A relationship between eustasy and biology is suggested by several periods during which great extinctions of marine organisms occurred, all coinciding with sudden decreases in global water depths. The detailed records provided by the deep sea cores show that Mesozoic climate changes were not gradual but were instead quite abrupt, probably much more so than one would expect if the chief cause had been, for example, the slow movements of the continents. Among the feasible causes of these abrupt climate events are periods with extensive volcanic activity or, even more dramatic, the collision of a comet or asteroid (a bolide) with Earth.

It has been calculated that an impacting bolide 10 km or more in diameter would have a dramatic effect on the planetary atmosphere. Typical impact velocities of such an object would be 20 km s^{-1} or more, releasing 60 million megatons of energy, about 4 thousand million times the energy of the Hiroshima atomic bomb, and creating a crater 150 km in diameter. The high rate of shock heating associated with such an impact would warm a substantial fraction of the atmosphere to 2000 to 3000°C. One immediate consequence is that the high concentrations of NO_x that would form would rapidly react with ozone and remove it from the atmosphere, thus allowing, at least initially, the penetration of high-energy ultraviolet radiation to the surface. A second consequence is that the nitric and nitrous acids produced from the NO_x would be incorporated into cloud and rain droplets and deposited on the land and sea as precipitation. The acidity of the precipitation would depend on the amount of acid in the atmosphere and on how fast it was taken up by the

Climates of the Past

droplets; estimates of these factors suggest that very high acidities might have been possible, a situation that could have caused the dissolution of the calcareous shells of marine life and the death of most land plants.

Changes in atmospheric chemistry would not be the only result of a bolide impact. Such an impact, if on land, would also create a global dust cloud sufficiently dense to block out the Sun's radiation for several months. The decrease in solar radiation would inhibit photosynthesis, resulting in a rapid die-off of plants and thus a loss of food for herbivores. The heat would ignite extensive wildfires, which would consume a large fraction of Earth's biomass. (In apparent support of this idea, large amounts of soot, suggesting worldwide forest and grass fires, have been detected in the sediments at several interepoch boundaries.) High levels of soot particles in the troposphere would have caused a severe cooling at Earth's surface, leading to what might be called a bolide winter. After the soot and dust settled, high concentrations of carbon dioxide, nitrous oxide, methane, and other species emitted by the fires and the rotting vegetation might remain in the atmosphere for centuries to millennia, leading eventually to a substantial climate warming due to an enhanced greenhouse effect.

More likely than a bolide collision with land is a bolide collision with water, since the oceans cover nearly three-quarters of the planet. Such an impact could volatilize carbonate-rich sediments into carbon dioxide, again leading to a substantially enhanced greenhouse effect.

The seeming implausibility of a bolide crashing into a planet was dramatically refuted in July of 1994, when 20 or so pieces of the disintegrating comet Shoemaker-Levy 9 slammed into Jupiter over a period of several days. The major disruptions produced in the atmosphere of the solar system's largest gaseous planet left little doubt that a similar collision with Earth could indeed occur—and that the effects could easily be catastrophic.

If we concede that bolide impacts with Earth might have occurred in the past, and recognize that widespread extinctions of much of the life on Earth are known to have occurred a number of times during the Paleozoic and Mesozoic eras and after, can we go on to demonstrate that the extinctions are related to bolide impacts? The extinction event that seems most likely to have been so caused occurred some 65 million years ago, at the Cretaceous-Tertiary geological boundary. In addition to the biological evidence of a sudden extinction of about 70% of Earth's

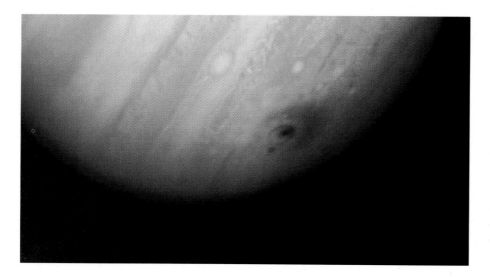

Jupiter shows the results of impacts of two fragments of Comet Shoemaker-Levy 9 on July 17–18, 1994. The dark thick outermost ring has a diameter of about the size of Earth.

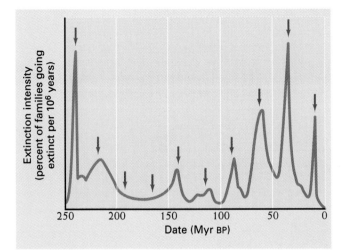

Species extinctions during the past 250 million years. The arrows are spaced 26 million years apart. The highest peaks represent loss of 50% or more of the observed families, but differences in species diversity in the fossil record make quantitative peak-to-peak comparisons problematic. These data are the basis of the theory of periodic extinctions in which a postulated solar companion star with a 26-million-year period, when nearing the point in its orbit closest to the Sun, perturbs comets orbiting outside the solar system and sends them into trajectories that intersect the orbit of Earth.

species, analyses of rocks deposited at that time reveal anomalously high levels of iridium, an element that is normally extremely scarce. The high iridium concentration at the interperiod boundary is global in extent and has been attributed by a number of researchers to the collision of an iridium-rich meteorite with Earth. In addition to the iridium evidence, the presence, in the same strata, of mineral glasses thought to form only under tremendous shock impact and of soot showing widespread fires indicates in completely different ways that a major bolide impact probably occurred.

The evidence to support other bolide impacts is less certain than that for the supposed impact at the Cretaceous-Tertiary boundary, but additional events are gradually becoming accepted as more likely than not. For example, some scientists have postulated that data such as those shown in the graph above suggest an impact cycle with a periodicity of about 26 million years. Evolutionary

biologists are thus starting to consider the possibility that the evolution of life on Earth has not been a continuous process but more like a series of bolide-caused restarts, each taking off from a different rung on the evolutionary ladder.

Cenozoic Climate: The Pre-Holocene

Uranium-lead dating of rocks, so crucial for studying the strata of the Mesozoic era, continues to be one of the primary techniques for dating the rocks of the pre-Holocene portion of the Cenozoic era, the period extending from 65 Myr to 10 kyr BP. A similar radiometric series, with potassium as parent and argon as daughter, is also used, although its utility is complicated by argon's being a gas and hence liable to escape from the rocks. Numerous strata analyzed by one or the other of these dating methods provide good signatures of the pre-Holocene climate. For example, they allow scientists to estimate the sea level at specific times in Earth's history by tracing the ^{18}O temperature record in fossils, with the amount of isotope enhancement indicating the degree to which ocean heating and expansion (or cooling and contraction) would have changed sea level height.

The forces responsible for the major climatic features seen during the pre-Holocene period, however, remain somewhat unclear. This was a time of reorganizations of Earth's land masses as the Pangaea supercontinent fragmented and plate tectonic forces slowly drove the separated continents toward the positions in which we find them today. This evolution may have had a significant influence on climate, as may the elevated levels of carbon dioxide that some data suggest. In any case, much of the Cenozoic era was rather warm (though not as warm as the late Mesozoic) until about 30 Myr BP, when an interval of glaciation began. This interglacial to glacial change could have resulted from modifications in ocean circulation, and, in fact, there is evidence in ocean sediment fossil records to suggest that this is what happened, although why such a change in circulation should have occurred at that time has not been convincingly explained.

We have stated that climate is driven by the influx of solar radiation to Earth, and have showed how the average intensity of that radiation is thought to have changed

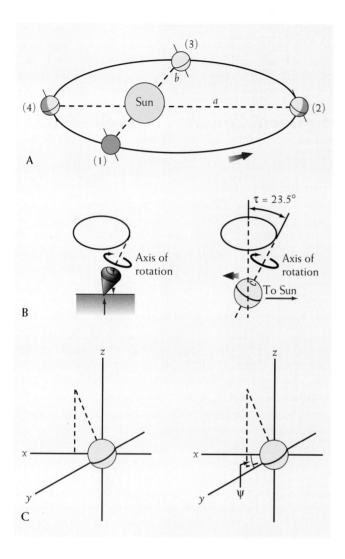

(A) The orientation of the axis of Earth in its orbit. In the course of one year, the orientation of the axis in space remains almost the same. Nevertheless, a planet has seasons when its equator is inclined relative to the plane of its orbit. For Earth, the obliquity (tilt) is 23.5° and the seasons are pronounced. Points 1 and 3 are those of the vernal and autumnal equinoxes, when, as Earth rotates about its axis, all latitudes receive twelve hours of sunlight. Three months after the vernal equinox, at the summer solstice of point 2, the northern hemisphere receives sunlight for more than half the day, the southern hemisphere less. The situation is reversed at the winter solstice, point 4, when the southern hemisphere receives sunlight for a greater portion of the day than does the northern. (B) Precession of a spinning top (left) and of the spinning Earth (right). The attraction of Earth upon the center of gravity of the inclined top would allow it to fall. This is prevented by the rotation of the top. The result of these competing forces is a conical motion of the top's axis of rotation. In the case of Earth, axial rotation prevents the equatorial bulge from assuming the orientation of the ecliptic. The result is a conical motion of the axis of rotation of Earth. (C) Precession in Earth's axis of rotation (left) the axis of rotation falls within the plane created by the orbital path of Earth about the Sun (x) and the orthogonal direction to that path (z); (right) shows the movement of the axis of rotation out of that plane by an angle ψ as a consequence of the gravitational attractive forces of the Sun, the Moon, and planets. (xy defines the plane of Earth's orbit.)

over the Sun's 4.6-billion-year history. On shorter time scales than billions of years, however, the radiation intensity has varied not because of the intrinsic evolution of the Sun, but because of gradual and predictable alterations in the orbit of Earth as it revolves about the Sun, rotates on its own axis, and is gravitationally perturbed by other planets. These alterations can be precisely determined by astronomical observations and calculations, and their time scale for change turns out to be tens of thousands of years—just the time scale appropriate for reflection in Cenozoic climate.

Three parameters describe the orbital variations of planets, including Earth. One is the degree of ellipticity of the orbit, or the eccentricity, equal to the ratio b/a in the top diagram on this page. The second is the angle of tilt between the axis of rotation of a planet and the planet–Sun orbit. Just as a spinning top wobbles around its angle of tilt, Earth and other planets wobble as they precess around their axis of rotation. The third orbital parameter is the orientation of the axis of rotation with respect to the orbital path. All of these factors evolve because of gravitational attractions for Earth by the Sun, the Moon, and the other planets. These have been computed, and as a consequence, it has been possible to correlate their magnitudes with climate records a long way back into the past.

The influence of orbital variations on the solar radiation received by Earth turns out to be neatly reflected in Earth's climate. A typical example is the situation of about 10,000 years ago, when the inclination of Earth's rotational axis relative to the plane of the Earth–Sun orbit was about 24.5° and the closest approach of the orbiting Earth to the Sun (the perihelion) occurred in July. Compared with the present situation—a 23.5° angle of inclination and a January perihelion—this past condition resulted in about 8% more solar radiation reaching the northern hemisphere at the summer solstice and about 8% less at the winter solstice. As we discussed in Chapter 2, the distribution and intensity of received radiation define the flow of the winds, the rates of precipitation, and other climatic conditions, so it can readily be seen that solar radiation changes of that magnitude would have had direct and noticeable impacts on climate.

The Deep Sea Drilling Program has recovered samples from many early- and mid-Cenozoic sediments around the world. These sediments have been dated by correlation with the global rock chronology and analyzed for, among other things, the ^{18}O content of the fossil shells. An example of the results, a composite plot for all the ma-

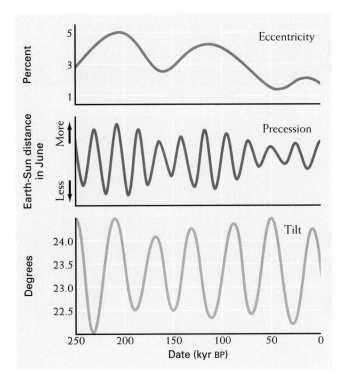

Long-term variations in three of Earth's orbital parameters that influence climate: eccentricity; precession; and tilt. The time scale is from 250 kyr BP to the present.

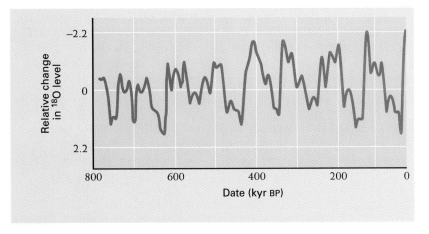

Relative changes in the ratio of ^{18}O to ^{16}O concentrations from deep sea cores. The data are plotted so that interglacial conditions appear near the top of the graph and glacial conditions appear near the bottom.

Climates of the Past

The U-shaped Jago River Valley in the Romanzof Arctic National Wildlife Refuge, Brooks Range, Alaska, carved by the advance and retreat of glaciers responding to changes in climate over millions of years.

jor ocean basins for the past million years, is shown on the previous page. Many fluctuations can be seen in these rather "noisy" data, but detailed mathematical analyses reveal glacial to interglacial cycles with periodicities of about 100,000, 41,000, 23,000, and 19,000 years. These periodicities all turn out to be correlated with the variations in solar irradiation that occur as a consequence of the progression of Earth's orbital parameters.

Just as ancient substances in the rocks permit us to decipher signals of ancient climates, so the use of ice cores enables us to study the climates of the more immediate past. The technique has been applied at locations including Antarctica and Greenland and on some mountain glaciers, where temperatures seldom or never exceed the freezing point of water. The procedure is to locate an appropriate ice mass, drill a core in such a way that the sample is retrieved in a minimally contaminated state, and return the core in its frozen state to the laboratory for study. The chemistry of the ice itself or of the soluble chemical species frozen into the ice is studied by slicing off a section of the ice corresponding to the desired age period, removing the contaminated outer surface, melting the re-

mainder under ultraclean conditions, and analyzing the resulting solution. If gas concentrations are to be measured, the uncontaminated portion of the ice sample is placed in a vacuum chamber and cracked; the gases are then sucked out and analyzed. If ionic concentrations from the frozen snow are to be determined, the melted sample is analyzed directly by solution chemistry techniques.

Atmospheric records in ice do not extend back in time indefinitely, because the lowest layers of ice gradually become so compressed by the weight of the new ice formed on top of them that they flow horizontally under pressure and their individuality and identity are lost. The oldest ice cores drilled and analyzed thus far have been dated to nearly 250 kyr BP in Greenland and about 500 kyr BP in Antarctica. The cores taken from these locations are about 2000 m long and about 10 cm in diameter. The core is dated by counting annual ice layers, much as a forester tells tree ages by the rings, and in the upper portions of the core the annual layers are distinct enough that annual ice layer counts have been done back to about 50 kyr BP. Reference dates are also established by evidence of known deposition events, such as layers containing radioactive

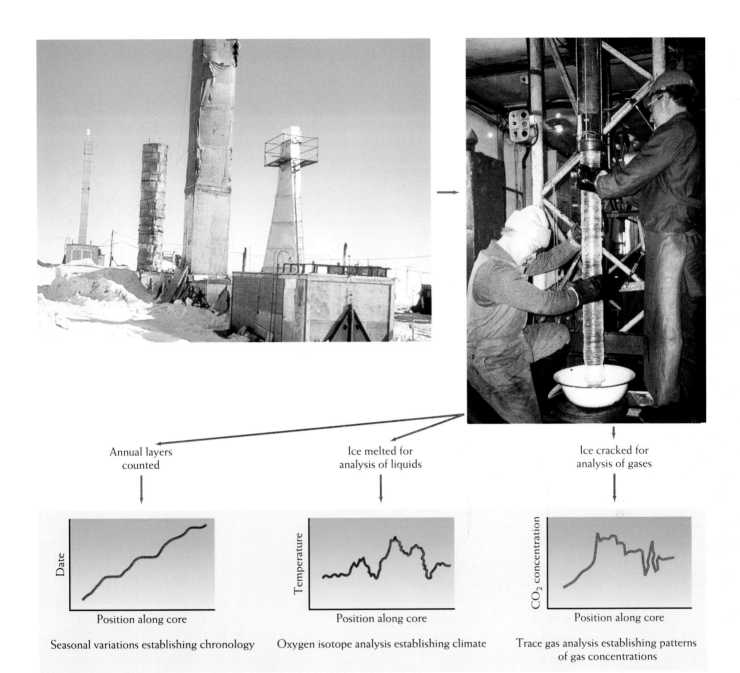

Annual layers
counted

Ice melted for
analysis of liquids

Ice cracked for
analysis of gases

Seasonal variations establishing chronology

Oxygen isotope analysis establishing climate

Trace gas analysis establishing patterns
of gas concentrations

(Top left) *The drill tower and associated structures at the former Soviet Union's Vostok, Antarctica facility.* (Top right) *Russian scientists removing the ice core from the drill string prior to its examination. The regions of annual surface thawing and recrystallization are clearly visible.* (Bottom) *Schematic diagrams of some of the information produced by the scientific analyses of ice cores.*

fallout from nuclear weapons tests or ash deposition from large volcanic eruptions. The oldest ice can be dated only by theoretical studies of the flow of the ice sheet floor when it is distorted by the pressure of the overlying ice. In general, as with rock, the older the sample, the less accurate the ice dating becomes. The bottom of the deepest datable Antarctic core is older than the bottom of the deepest Greenland core because the annual snowfall is lighter in Antarctica than in Greenland; a corollary of this fact is that the temporal resolution of data from the Antarctic ice cores is lower than is that for the Greenland cores.

The snow that fell and ultimately became the polar ice cores consisted of water evaporated from land and ocean, and its oxygen isotope content thus reflects the temperature at the time of evaporation and deposition. The longest and most carefully studied record of polar-ice climate histories published thus far is from the ice core drilled in Antarctica from 1980 to 1985 at Russia's Vostok station and analyzed by a team of French and Russian scientists. The tale it tells is one of frequent change in climate over the past 150,000 years, not stability. The highest temperatures during that period occurred about 130 kyr BP. During most of the time since then, the temperature has been several degrees lower but has fluctuated substantially. The major peaks and troughs in the record are associated in part with the changes in Earth's orbit that we mentioned above. The causes of the smaller fluctuations, still a degree or so in magnitude, are not yet understood in detail, but this is not surprising considering that we are dealing with a complex climate system that encompasses many reinforcing and opposing forces.

Cenozoic Climate: The Holocene

By far the most extensive record of past climates that we have is for our own epoch, known as the Holocene (the period since the last great ice age 10 kyr BP), and the few thousand years immediately preceding it. Many different techniques have proved useful in deriving the detailed climatic history of various locales and integrating that information into regional and global assessments of Holocene climate.

A valuable tool for dating substances on Earth that have lifetimes much shorter than those of rocks is one

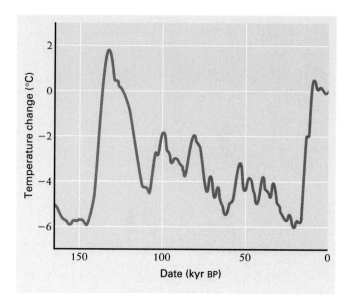

The temperature record from the Vostok ice core, Antarctica, for the late Cenozoic.

based on the isotope carbon-14 (^{14}C). This isotope of carbon is continually created in the upper atmosphere by the interaction of nitrogen atoms with high-energy cosmic ray neutrons from outside the solar system:

$$^{14}_{7}N + ^{1}_{0}n \rightarrow ^{14}_{6}C + ^{1}_{1}H \qquad (4.5)$$

(In the notation, the atomic weight is given at the upper left, the number of protons at the lower left.) The carbon atoms form carbon monoxide by combining with molecular oxygen,

$$^{14}_{6}C + O_2 \rightarrow ^{14}_{6}CO + O \qquad (4.6)$$

are oxidized to carbon dioxide by reaction with the hydroxyl radical,

$$^{14}_{6}CO + HO\cdot \rightarrow ^{14}_{6}CO_2 + H\cdot \qquad (4.7)$$

and eventually decay by emitting a high-energy electron (a beta particle):

$$^{14}_{6}CO_2 \rightarrow ^{14}_{7}NO_2 + \beta \qquad (4.8)$$

On average, the lifetime of the carbon isotopes before they decay is about 5730 years. Before the decay, however, the radioactive carbon dioxide may be utilized by growing vegetation in the same way as the more common carbon dioxide containing carbon-12 (^{12}C). As a consequence, every living thing contains both of those isotopes of carbon, and the ^{14}C to ^{12}C isotope ratio in anything that was once alive reveals the time of the organism's death. Alternatively, if independent dating techniques are available (for example, a tree's age can be determined by counting its rings), then the ratio measurement permits an analysis to be made of the amount of ^{14}C present in the atmosphere at a known time in the past, thus providing a marker of solar activity.

The sediments of lakes and wetlands are also potentially rich sources of information about past environments, not only within the lakes or wetlands themselves, but also in the surrounding terrain. The sediments have two sources: internal processes within the lake itself and particles brought to the lake by wind or by runoff from the land. Sediments that may be suitable for scientific studies must be carefully dated (a variety of methods are available), and then some physical or chemical characteristic of the sediment may be selected and evaluated in each layer of the sediment core. The property chosen must be one that does not undergo change after it is fixed into position; among such properties are the presence of volcanic minerals, grains of specific pollens, and microfossils.

An example of information available from sediment cores is the pollen data collected from a bog in northern Minnesota, shown in the figure on the following page. Scientists have traced the local history of pollen from 14 plant taxa back about 11,000 years in the bog sediments. In the oldest layers studied, spruce pollen is dominant, implying that the climate at that time was quite cool. Soon after, pine became dominant, a change indicating a significant warming. The pine dominance gave way to oak at about 8.5 kyr BP, a pattern typical of drier climates. Birch coexisted with oak for a period of about 2000 years. Finally, this dual dominance yielded to the current overwhelming abundance of pine. Sediment cores from this and other sites have provided similar information about the presence of grasses and other vegetative species, all indicators of particular temperature ranges and climates.

Trees are the oldest living things on Earth, some existing bristlecone pines having been venerable when Jesus lived. They are especially rich repositories of atmospheric information for a number of reasons: they interact with the atmosphere (and retain signatures of that interaction); they reflect some of the properties of the climate in which they grow; their age provides long, continuous records; and tree records can be very accurately dated, generally to within a specific year. These properties are all reflected in the rings produced by trees as they undergo their annual cycle of growth. Tree-ring records may be obtained from dead but well-preserved trees, from tree stumps, and from cores taken from living trees, the latter establishing an extensive and continuing climatological record.

The most comprehensive look into paleoclimates has been the work of the Cooperative Holocene Mapping Project (COHMAP). The purpose of this international research group is to understand the climates of the past 20,000 years, its main goals being to assemble global sets of paleoclimatic data, calibrate the data in climatic terms, and use the data to test the ability of global circulation models to simulate past climates. The results of these efforts can be seen in the maps on pages 84–85.

COHMAP primarily used data on pollen, lake levels, and marine plankton to perform its assessment, although complementary tree ring and rock structure evidence was incorporated as well. The group's principal finding is that the driving force behind much of the long-term climatic change, not just the changes in planetary temperature, was the variation in the solar radiation received at Earth's surface as a result of the changing orientation of Earth's axis in orbit about the Sun. The orientation at 18 kyr BP was similar to that of today. In contrast, between 15 and 9 kyr BP, the Sun–Earth distance decreased at the time of summer in the northern hemisphere and the axial tilt increased; as a result, seasonality was enhanced more in the northern hemisphere and less in the southern hemisphere than is the case at present. By 9 kyr BP, the average solar radiation over the northern hemisphere was 8% higher in July and 8% lower in January than is the case today. Since that time, these levels have gradually approached modern values.

The major effect of the radiation intensity change between 12 and 6 kyr BP was to enhance the temperature contrast between ocean and land, causing stronger evaporation from the oceans and increasing weather-cell formation and precipitation over the continents. This pic-

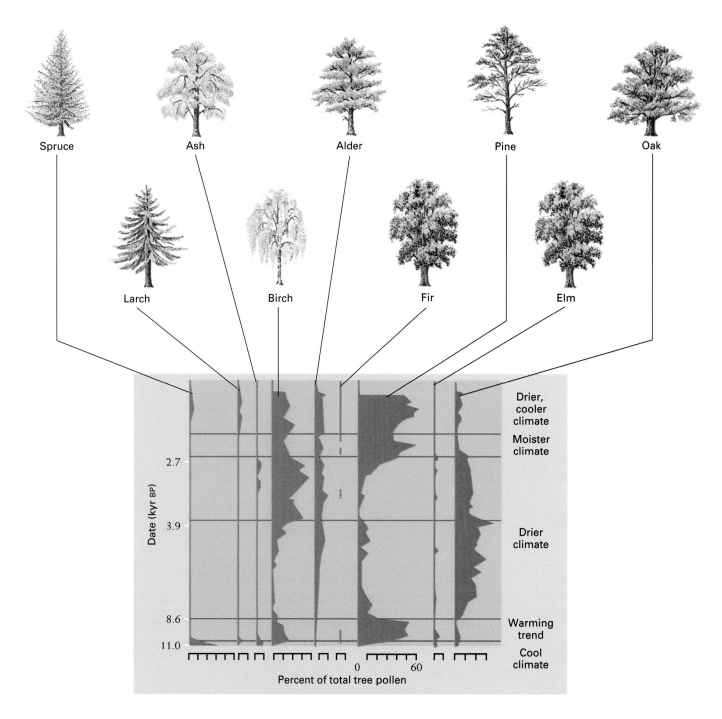

Pollen diagram for Bog D in Itasca State Park, northwestern Minnesota, over the period from 11 kyr BP to the present.

Winter Landscape, *painted in 1601 by Peter Brueghel the Younger, at a time when the Little Ice Age temperatures regularly froze the canals of Holland, now generally ice-free the year around.*

ture is supported by the pollen, lake-level, and fossil records and has been reproduced by computerized meteorological models of atmospheric circulation and precipitation, which we will discuss in Chapter 6. Assuming that these changes in the rock and fossil records are typical of the kinds of fluctuations that occur in the natural climate system, much can be learned by comparing the present-day situation with those of the past.

In today's climate in the Northern Hemisphere, permanent land ice occurs mainly in Greenland but not in North America or Eurasia, whereas sea ice covers most of the Arctic Ocean. Spruce forests occupy much of the high-latitude regions below the land-ice latitudes, and oak is fa-

vored in the temperate latitudes. Low levels of precipitation have created deserts in several areas. Subtropical conditions occur at the latitudes of Florida and northern Africa. In contrast, the climate of 18,000 years ago in the far north was characterized by large ice sheets on land and substantial sea ice in the North Atlantic. Spruce and oak forests were absent or sparse in Europe, and the Mediterranean lowlands were without trees. Spruce dominated the forests of the North American Midwest. In the Southwest, lake levels were high and woodlands were abundant. As time progressed, the increasingly warm climate pushed many of the forests northward. These reached their maximum latitudes by about 6 kyr BP. Since then, summer

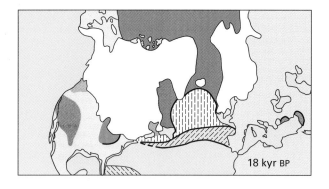

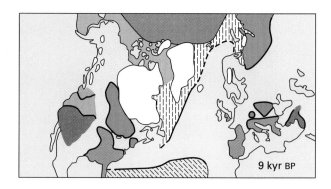

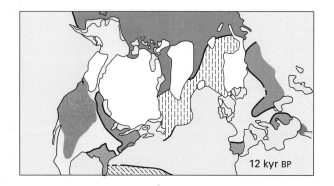

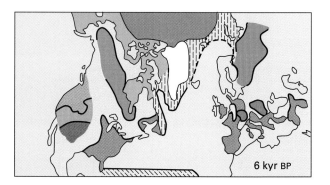

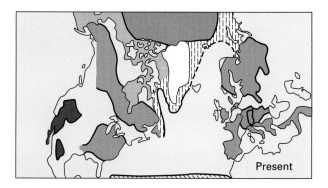

Changes in the atmosphere, geosphere, and biosphere that accompanied the transition from glacial to interglacial conditions during the past 18,000 years, as illustrated by geological and paleoecological evidence for eastern North America and Europe (on this page) and Africa, Asia, and Australia (on the facing page). The individual panels show the extent of ice sheets and of year-round and winter-only sea ice; from 18 to 9 kyr BP they show the broadened land areas resulting from lowered sea level. The distributions of oak and spruce as inferred from pollen data are shown for eastern North America and Europe. Moisture status relative to the present is displayed for all areas from 18 to 6 kyr BP, as is the present region where annual precipitation is less than 300 mm.

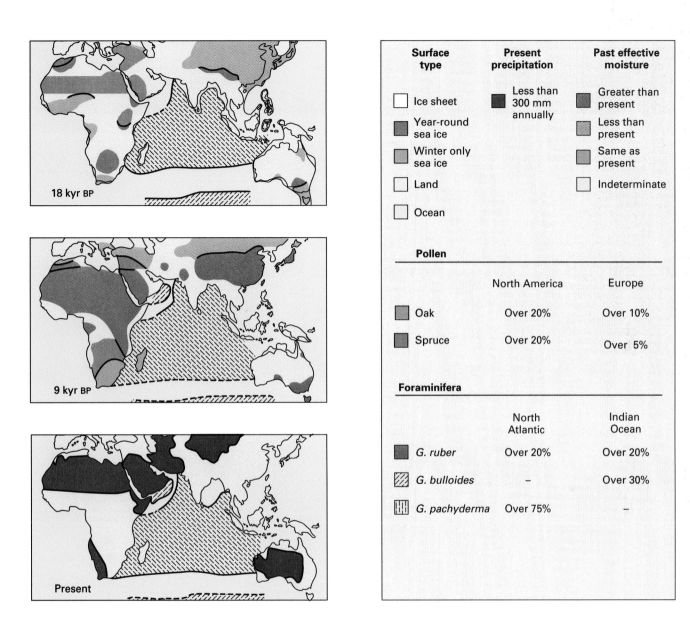

Surface type	Present precipitation	Past effective moisture
☐ Ice sheet	■ Less than 300 mm annually	■ Greater than present
■ Year-round sea ice		■ Less than present
■ Winter only sea ice		■ Same as present
☐ Land		☐ Indeterminate
☐ Ocean		

Pollen

	North America	Europe
■ Oak	Over 20%	Over 10%
■ Spruce	Over 20%	Over 5%

Foraminifera

	North Atlantic	Indian Ocean
■ *G. ruber*	Over 20%	Over 20%
▨ *G. bulloides*	–	Over 30%
▦ *G. pachyderma*	Over 75%	–

temperatures have decreased somewhat, and the southern limits of several of the tree species have begun once more to shift toward the equator.

Summer monsoons—periods of strong sea-to-land winds and heavy precipitation—dominate the present-day climate in Africa, Asia, and Australia. Great disparities in precipitation exist over these continents, with equatorial rain forest in some areas and dry desert regions in others. The data for 18,000 years ago show much smaller areas of forest, an indication that the Mediterranean and its environs were cooler and perhaps drier than today. Later, in the period 12 to 6 kyr BP, greater precipitation, probably caused by the gradual formation of the strong monsoon pattern, resulted in increased vegetation. Somewhat drier conditions have since followed, as solar radiation gradually decreased to present values in accordance with the changes in Sun–Earth orbital and angular relationships that we discussed above.

The COHMAP group used computer models to test their ideas about the probable climate conditions that existed during the Holocene. Specifically, they inserted their estimates of changes in solar insolation, mountain and ice sheet topography, atmospheric gas concentrations, sea surface temperatures, snow cover, and soil moisture into a modern climate model and compared the computed results with the paleoclimatic data. Not only did the model produce climate results that seemed realistic, but it correctly predicted such features as the location of jet streams, the

General trends in global temperature on various time scales ranging from decades to hundreds of thousands of years. The bars at the right indicate the temperature range over the time illustrated by each graph. On the bottom three displays, the shaded area indicates the time segment that is represented in the display immediately above. The data reconstructions are based primarily on the following records: (top) instrumental data, (second from top) historical information, (second from bottom) pollen data and alpine glacier progression and retrogression, and (bottom) marine plankton.

strength of the atmospheric circulation, and the location of deserts at different times during the Holocene era.

The central lesson that climatologists have learned from the COHMAP study and other similar efforts is the significant extent to which many aspects of climate change are a natural result of orbitally induced variations in the magnitude and seasonality of solar radiation. These cause periodic temperature changes that modify not only the glacial ice sheets but also the sea-to-land circulation patterns, precipitation intensity and seasonality, and patterns of plant growth and decline (and hence the response of the animal kingdom). Not only is climate *inherently* variable because of these orbital changes, but the ways in which it varies can be generally understood as consequences of the most basic aspects of Earth's astronomical and surface properties. As we consider in later chapters the potential changes to atmosphere and climate resulting from human activities and impulsive natural events, it will be important to remember that they are superimposed on this background of orbit-induced variability.

A Synopsis of Climate Histories

It is truly startling to realize that we possess a general idea of Earth's changing climate almost from the time of the planet's beginning more than 4 Gyr BP. Of course, our knowledge is more complete for later periods than for earlier ones, because more information is available; and for the past 250,000 years the climate record, shown in the series of graphs on the facing page, is moderately well detailed. For the past 20,000 years, the records are so complete that we know not only the typical temperature and precipitation patterns but also features as precise as the geographical ranges of specific species of trees.

Despite our fascination with very ancient climates, we are doubtless most interested in results for more recent periods, which show that Earth's climate has changed from one characterized by extreme glaciation to a warm interglacial condition in the past 20,000 years. Computer simulations suggest that midway through this time span,

changes in atmospheric circulation increased monsoon precipitation and warmed the continental interiors. One result was the creation, in the northern latitudes, of a climate that was comfortable for humans, and indeed for living organisms in general. This no doubt contributed to the expansion of human activity in those regions and was a major determining factor in the history of civilization.

The past thousand years, too, have seen significant changes in global climate, though the temperature range over that period has been much smaller than over earlier periods discussed above. The most striking features in the graphs showing average temperatures for the immediate past millennium are the sharp cooling between about AD 1400 and AD 1650, during the so-called Little Ice Age, when the Baltic Sea and rivers such as the Thames in England and the Tagus in Spain, currently ice-free the year around, were regularly frozen several inches thick. It has been proposed that the climatological character of this period was related to a significant reduction in magnetic activity at the surface of the Sun, because sunspots were virtually absent during that period. No satisfactory hypothesis has yet been developed, however, to explain why this aspect of solar activity should affect Earth's climate.

For the past century, we can rely on direct instrumental measures of temperature. During that period, and especially during the past 20 to 30 years, a definite warming of the planet has occurred; that point is not contested, though its cause is a topic of vigorous debate. Some experts argue for natural variability, and certainly Earth's history is one of rich climate variation. Others argue for the greenhouse effect, and certainly large quantities of several greenhouse gases that could contribute to such an effect, such as carbon dioxide, nitrous oxide, and methane, have been added to the atmosphere by humanity's actions. The issue cannot be decided with certainty at this point, but it can be discussed, and rough odds can be assigned to the probability of various outcomes, as we will attempt to do later in this book. Before we can look toward the future, however, we must examine in the next chapter what we know of past changes in atmospheric composition (one of the drivers of climate), much as we have done in this chapter for climate itself.

*Patrick Zimmerman of the U.S. National Center for Atmospheric Research escapes to safety
with a flask containing a sample of air from burning grass and brush in Brazil's Amazon
region. Biomass burning is commonly practiced by farmers in tropical countries for land clearing
and pesticide control; its effects on the atmosphere can be detected half a world away.*

Changing Chemistry

5

The first rule of intelligent tinkering is to save all the parts.
—Aldo Leopold

One explanation proposed for the weak sun paradox—in which a Sun fainter than today's nonetheless maintained a temperature above the freezing point of water on the young Earth—is that the levels of atmospheric carbon dioxide may have been substantially higher at that time. This conjecture reminds us that atmospheric composition may be no more immutable than climate is. Earth scientists can deduce past climates from records in rocks, ice, and trees. Can they deduce past atmospheric compositions as well? In this chapter we will search for the answer to that question, both for the distant past, when natural processes were the only possible agents of change, and for the recent past, when humans appeared and began to influence their environment in various unforeseen ways. We will also explore the current effects of human activities on atmospheric composition. Is our "tinkering" destroying certain parts of the Earth system? Are we ignoring the advice of naturalist and conservationist Aldo Leopold?

It seems almost inconceivable that very much could be learned about the chemistry of Earth's atmosphere further back in time than the half century or so during which modern analytical instruments have been developed and used. But, in fact, Earth scientists have discovered a number of techniques for studying historical chemistry. One such approach is based on the recognition that chemical interactions occur between atmospheric constituents and the surfaces with which they come into contact. If these interactions are understood, and if the surfaces are preserved and retain the signature of the interaction, then ancient surfaces can be useful in establishing past atmospheric characteristics. The most fruitful data of this type come from the rock record, because it is known that certain dated rocks could have been formed only under specific atmospheric conditions. Even though this technique cannot be used to derive information about trace constituents of the past, it does confirm the atmospheric presence in the atmosphere of ancient Earth of the major constituents molecular oxygen and carbon dioxide.

The earliest known evidence for any atmospheric species is that for carbon dioxide, gathered by Robert Berner of Yale University. Berner calculates carbon dioxide levels using several categories of information: the spreading rates deduced for ancient sea floors and the inferred outgassing of carbon dioxide from molten rock as it flowed from Earth's interior onto the ocean floor at midocean ridges; inferred rates of carbon dioxide release by the weathering of carbonate rocks; and amount of vegetative land cover available to take up carbon dioxide in photosynthesis. The derived concentrations, shown in the graphs on the facing page, are by no means certain, but the result is not unreasonable: carbon dioxide levels 15 to 20 times those of today were present at about the beginning of the Cambrian period, when life on Earth was evolving rapidly, and carbon dioxide concentrations declined to near those of today by about 300 Myr BP, when rock weathering was inhibited during the major Paleozoic glaciation. The oldest *direct* information about atmospheric composition comes from Mesozoic fossils recovered from deep sea sediment cores. Photosynthetic plankton use dissolved carbon dioxide to form chlorophyll, a portion of which is eventually deposited in the ocean sediments. The plankton prefer (by a very slight margin) to utilize $^{12}CO_2$ rather than $^{13}CO_2$ when both are readily available in the

surrounding water (^{13}C is the heavier and far less common isotope). If dissolved carbon dioxide were scarce, the plankton would have been forced to use whatever they could get, incorporating both isotopes of carbon without preference. If dissolved carbon dioxide were abundant, the plankton would have preferentially chosen $^{12}CO_2$. Thus, periods of carbon dioxide abundance correspond to periods in the sediment record when the planktonic fossils are abnormally rich in $^{12}CO_2$.

The relative abundance of dissolved carbon dioxide in seawater is, of course, related to the concentration of carbon dioxide in the atmosphere. Consequently, the relative abundances of the carbon isotopes in dated sedimentary fossils can be used to deduce the atmospheric concentrations of carbon dioxide over time. The general pattern of the data suggests that the atmospheric carbon dioxide concentrations of 150 Myr BP were about 3 times higher than at present, more or less consistent with Berner's results.

For periods more recent than 500 kyr BP, there is an electrifying difference in the quality of our knowledge about the atmosphere's composition, thanks to the availability of actual samples of ancient atmospheric air that can be analyzed by modern analytical techniques. These samples are extracted from cores of polar ice, which, as we saw in Chapter 4, forms from the snow that falls each year on glaciers and then consolidates. Particles and soluble gases incorporated along with the snowflakes or absorbed at the snow surface remain in the layer corresponding to the time of deposition.

Ice core techniques are particularly appropriate for investigating species that are relatively unreactive, because the possible complication of losing those species through chemical reaction during or after freezing is minimized. Species of this type are long lived and thus well mixed in the atmosphere, so the ice sample provides an approximation of their average atmospheric concentrations worldwide. Ice core concentrations of reactive and water-soluble gases, as well as aerosol particles, are more reflective of local or regional, rather than global, conditions, although in some cases the substances may have been transported to the ice from quite distant locations. Thus, although the deep ice cores reflect the particular characteristics of the continent on which they formed, they also serve as an integrated record of the global at-

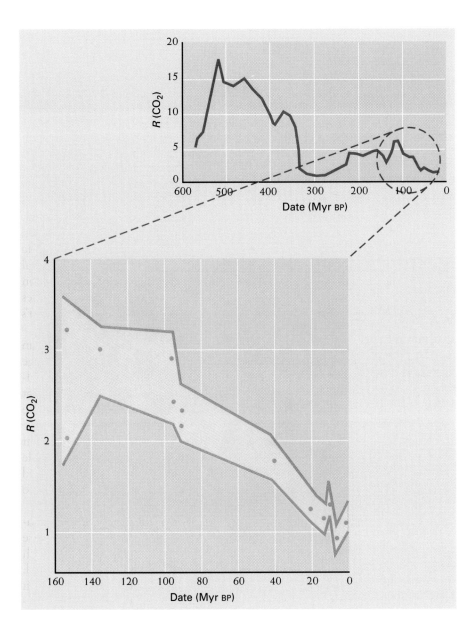

mosphere for gases with lifetimes of at least several years, because that is the approximate time scale for effective mixing between the air in the different hemispheres.

The oldest sample we have of the chemical composition of the atmosphere is from the Vostok Antarctic ice core. Drilled by scientists from the former Soviet Union and analyzed in collaboration with French colleagues, its deepest segment was laid down nearly 500 kyr BP. This is the ice core that provided the record of atmospheric temperature that we illustrated in the previous chapter.

When the gases from the ice are extracted and analyzed for carbon dioxide content, the correlation between global temperature and carbon dioxide concentration is striking. The carbon dioxide pattern mirrors the warm interglacial temperature peak at 130 to 140 kyr BP, the gradual decline to a glacial epoch at 20 kyr BP (note how long the glacial episode lasted in comparison to the warm interglacial one), and the rapid warming from that point to the present day. It should be noted that the highest carbon dioxide concentrations in history prior to very recent times are those of 130 to 140 kyr BP, toward the end of the Pleistocene epoch; they peak at over 290 ppmv and decrease over a period of 4000 years to about 230 ppmv, the average rate of change being about 1 ppmv per 100 years. (The present average carbon dioxide concentration in the remote atmosphere of the Antarctic is about 360 ppmv.) Preindustrial carbon dioxide levels during the present interglacial interval, the Holocene epoch, were in the range of 270 to 280 ppmv, while carbon dioxide concentrations of the most recent ice age were about 80 to 90 ppmv lower. An interesting perspective is gained when we consider that the present rate of carbon dioxide increase as a consequence of fossil fuel combustion is larger than 1 ppmv *per year*. In other words, the modern rate of change of this major atmospheric constituent of crucial importance to the biosphere and climate is about 100 times faster than the highest average rate that had occurred prior to the present century, and it is occurring at a time when the absolute concentration of carbon dioxide is higher than it has been at any time during at least the past several hundred thousand years.

Because the concentrations of other trace gases in the atmosphere are so much lower than that of carbon dioxide, their presence is not as easily detected in small samples of ice, and the records of their concentrations over time are much sparser, particularly for the oldest, most compressed layers of ice. Of the data that are available, perhaps the most interesting are those of methane concentrations in the Vostok ice core. Like carbon dioxide, methane appears to have undergone changes in concentration that parallel the fluctuations of temperature, as shown in the graphs to the right. This pattern reflects the gain or loss of wetlands due to changing ice cover, as well as the temperature-sensitive growth rates of the methane-emitting bacteria that inhabit the wetlands.

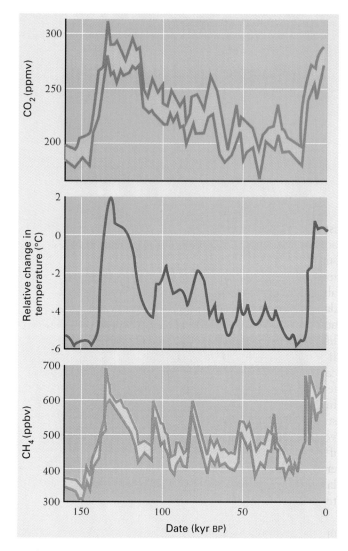

Carbon dioxide, oxygen-isotope temperature, and methane historical records from the Vostok ice core in Antarctica.

Ice contains a variety of trace constituents in addition to gases. Among those that are easiest to identify are sulfate (SO_4^{2-}) and nitrate (NO_3^-) ions, which represent material that was incorporated into rain, snow, or hail before it fell to the ground, solubilized ions from aerosol particles, and occasional major injections of ions from particles produced by volcanic eruptions. Analysis of ice from Greenland shows that ionic nitrate and sulfate levels were stable over nearly the entire time period represented by the ice cores extracted there; this stability was overthrown in each case by a very sharp increase during the present century. For each of these species, the increase can be attributed to emissions from fossil fuel burning, in combination with long-range air flows from industrial regions to the Arctic. The Antarctic, on the other hand, is so far from combustion sources that it receives little ionic sulfate or nitrate pollution except from volcanic eruptions, marine sources of biogenic sulfur, and deposition from the stratosphere.

Another measurement reflecting the chemistry of ancient precipitation is that of ice core acidity, a characteristic that largely corresponds to the concentrations of the acid-related sulfate and chloride (Cl^-) ions. Acidity measurements can be used in combination with the inorganic ion concentrations to derive considerable information about ambient chemical conditions. Together with the ion data, they have proved particularly useful as signposts of ancient volcanic eruptions, since most volcanoes are prolific emitters of sulfur and chlorine, as graphed on page 94. If the date of an eruption is already known from historical records, as is the case for many such events in the last few hundred years, the timing and intensity of the signal in the ice can be used to confirm ice core dating and to study changes in atmospheric transport efficiencies from volcano to ice sheet over long periods of time. Conversely, if written histories do not exist or make no mention of an eruption whose occurrence is indicated by the acidity of a layer of ancient ice, the ice record helps expand the historical data available to volcanologists.

Like ice cores from the polar caps, ice cores from the upper layers of midlatitude glaciers are another productive source of recent climate information, particularly concerning temperature and amount of precipitation. For example, temperature data obtained from a Peruvian ice core whose layers date from 1600 to the present agree well with independent temperature histories derived for the northern hemisphere, confirming that the Little Ice Age of 1400 to 1650 and the climatic impact of the 1815 volcanic eruption in Tambora, Indonesia, were global in extent and demonstrating that some midlatitude glacial records, too, can play important roles in studies of historical climate and air quality.

Replaying an Earth-atmosphere interaction that has occurred since time began, Mt. Pinatubo in The Philippines erupts on June 17, 1991, injecting sulfur gases and dust into the stratosphere and causing two years of moderate global cooling.

Human Influences on the Atmosphere

Throughout most of history, the number of humans on Earth was quite modest, and their impact on the landscape negligible. With the passage of time and the growth

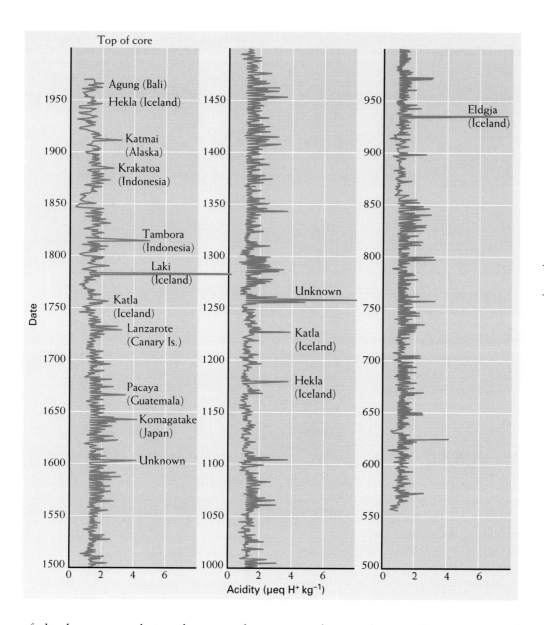

Top of core

The mean acidity of annual layers from AD 1972 to AD 553 in an ice core from central Greenland. Acidities above the background; of $1.2 \pm 1.2 \ \mu eq \ H^+ \ kg^{-1}$ ice, are caused by the fallout of volcanic acids, mainly H_2SO_4, from eruptions north of 20°S latitude. The ice core is dated with an incertinty of ± 1 year in the past 900 years, increasing to ± 3 years at AD 553. This precision makes it possible to identify fallout from several large volcanic eruptions known from historical sources, and to provide an accurate date for the Icelandic Eldgja eruption shortly after the settlement of Iceland in AD 930.

of the human population, however, the situation has changed. History suggests that large-scale environmental disruption by human beings began more than 2000 years ago with slash-and-burn agriculture and other exploitation of soils and forest resources in China, Mesopotamia, Egypt, Roman North Africa, and elsewhere. While resulting in noticeable local pollution, these activities had little effect on the global environment; nor did the use of wood (and

later coal) for residential heating have much global impact. But then, near the middle of the eighteenth century, came the Industrial Revolution.

This revolution in economic activity and lifestyle was powered by coal, the energy source for the newly invented steam engines and, as a result, for the vast increase in metals production. The abundant availability of iron permitted the manufacture of high-quality tools and machines and the building of improved bridges and ships. Agriculture, too, became mechanized, spurring the transformation from agrarian life to urban life for a sizeable portion of Earth's population. The Industrial Revolution was particularly successful, first in Europe and then in North America, in societies that were able to bring together within a small geographical region the necessary scientific and technical knowledge, substantial quantities of energy and capital, and a large work force.

Progressive industrialization increased the demand for metals other than iron, and techniques were developed to extract them from their ores. Copper became particularly important because of its high electrical conductivity and was produced in large quantity in the latter half of the nineteenth century. In recent decades, the extraction of zinc, aluminum, and other utilitarian metals from their ores has become common.

The advances of the Industrial Revolution have exacted a heavy toll on the environment. For example, of the mass of materials extracted from the ground for the production of useful metals, more than 90% are discarded as waste by-products. Concomitantly, the revolution brought the first large-scale anthropogenic emissions of gases and particles into the atmosphere, changing the quality of the very air people breathed. Coal use alone results in diverse and voluminous emissions. In fact, coal is now the most environmentally detrimental of energy sources, in terms of negative atmospheric impacts; these will almost certainly grow with time, because the world's reserves of coal are very large in comparison with those of petroleum and natural gas and because coal is generally the least expensive reliable source of energy. Thus, coal combustion is one of the most serious obstacles to the long-term sustainable development of the biosphere. In the future, governments may well require that coal use be preceded by a cleaning step to remove sulfur and other trace constituents. If so, the cost advantage coal now possesses over other forms of fuel will diminish or disappear.

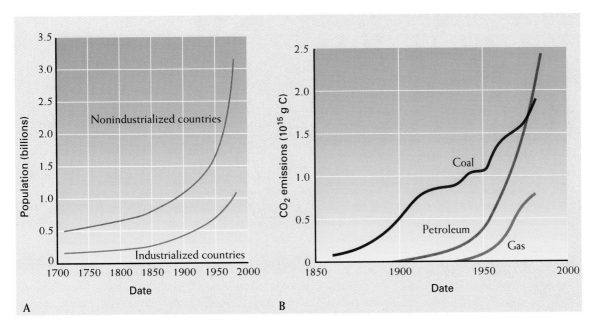

(A) The human population over the past several centuries. (B) The world use of coal, petroleum, and natural gas from 1860 to 1980, as measured by the emission of CO_2 to the atmosphere, expressed as petagrams $(10^{15}g)$ of carbon per year.

Throughout the eighteenth and nineteenth centuries, smokestacks such as those in this English painting were viewed as objects of civic pride.

While the use of petroleum as an energy source was negligible before the twentieth century, its consumption, especially by motor vehicles, has grown rapidly ever since. The most dramatic benchmark of its rise to energy prominence occurred in about 1975, when petroleum combustion passed coal combustion as a source of carbon dioxide. Petroleum dominance will probably persist for the next century, until world petroleum reserves become depleted to the extent that petroleum can no longer be a major energy source.

The third most common fossil-fuel energy source is natural gas, which is mostly methane. Little used prior to the first quarter of the twentieth century, its combustion for the generation of electrical energy has risen rapidly since that time. Natural gas tends to burn much more cleanly than coal or petroleum, with carbon dioxide being the only emittant of concern. Like the supply of petroleum, Earth's supply of recoverable natural gas is quite limited. It may be largely used up by the middle of

the next century, thereafter ceasing to be an important factor in atmospheric impact assessments.

The most diverse group of emissions that has entered the atmosphere comes from the wide variety of industrial processes and subsequent product use for which fuel combustion is not significant. Three classes of these emittants merit special attention. The first is the family of chlorofluorocarbon compounds, which have been the subject of recent international agreements intended to prohibit their emissions by the end of the century. A second emittant of note is the carbon dioxide arising from the conversion of limestone into lime to make cement. (Cement manufacture currently is responsible for about 2% of all anthropogenic carbon dioxide emissions.) The third are tiny particles, which play major roles in reducing visibility, can be harmful to human health, and serve as a platform for various chemical transformations of gaseous pollutants.

Coincident with the gradual appearance of these industrial emissions were other atmospheric changes resulting from an evolving Agricultural Revolution. Agriculture developed rapidly in many parts of the world during the early and mid-twentieth century, driven by population growth and by human immigration to areas previously sparsely populated and developed. In these regions, huge tracts of land were, and are still being, deforested to provide farmland. Rapid cycling of resources between the atmosphere and the ground also resulted from the combustion of biomass, especially dry grasses, chaff, and other agricultural by-products. This periodic burning continues, principally in the tropics during the dry season, to clear forest land for agriculture and grazing, to maintain soil productivity, to get rid of insects and other pests, to burn agricultural wastes and dry savanna grasses, and, of course, for heating and cooking. In addition to carbon dioxide, a whole suite of chemically reactive gases and particulate species are emitted as a consequence. The net effect of the agricultural and industrial activities has been a steady rise in the atmospheric concentration of carbon dioxide, as shown in the graph on the right.

Agriculture and the atmosphere also interact through the digestive processes of ruminants, in which methane is produced by fermentation of grasses. The number and

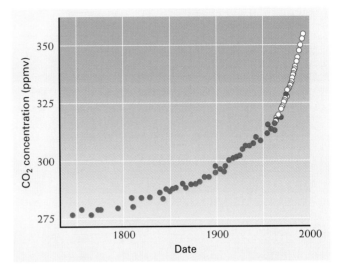

Carbon dioxide concentrations at remote global sites from 1750 through 1989. Solid circles represent data from an ice core drilled at Siple Station, Antarctica; open circles represent data acquired in situ by modern analytical techniques at Mauna Loa, Hawaii.

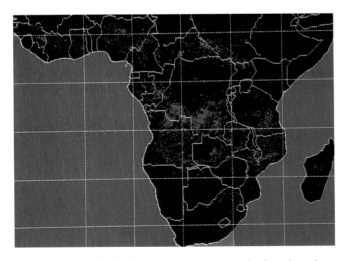

A map constructed of nighttime satellite images of Africa throughout 1987, with city lights removed. Each red dot indicates a biomass burning fire, which generally occurred during the dry season at the location where the fire took place.

mass of undomesticated animals on Earth has almost certainly decreased in recent times because of the steady expansion of humans. The effects of this decrease on atmospheric emissions of methane and ammonia, however, are more than compensated for—in fact, are overwhelmed by—the growth in domestic animal populations and the concomitant conversion of natural lands, especially forests, into pasture and intensely cultivated agricultural lands. Since about 1800, the number of domestic animals has increased dramatically, with proportional increases in the emissions of methane and ammonia. Because the diets of more highly industrialized societies contain more meat, the rapid increase in the number of domestic animals can be expected to continue as long as the supply of grain does not limit that increase. Although difficult to quantify in detail, the fluxes from the undomesticated animal kingdom appear to be quite small compared to those for other methane sources, but those from domestic ruminant animals are very significant, accounting for some 15% of the total methane emitted annually.

The effect of modern industrial, agricultural, and residential activities on atmospheric methane is easily seen

Cattle feedlots in Texas. The number of cattle on Earth increases every year, and some fifteen percent of all the emissions of methane, a greenhouse gas, occur as a consequence of the digestive processes of ruminants.

in the historical data. Evidence from ancient ice samples reveals that methane's abundance was essentially constant, at about 700 ppbv, until about 200 years ago. After that date, atmospheric methane began the relentless increase shown in the graph to the right. Ice core determinations of nitrous oxide and other gases resulting from human activity mirror this as well.

Thus, ice core data provide a surprisingly informative picture of several of the environment's trace species over the past few centuries. Depending on the lifetime of the substance undergoing study, the information can be global, regional, or local in scope. Nevertheless, in all cases that scientists have analyzed, the general picture is the same: relative constancy in concentrations until a few hundred years ago, followed by increases, at various rates, in the concentrations of nearly all the measured constituents that have anthropogenic sources—first for agriculturally related species, then for species in industrial emissions. Although we can never hope to construct a complete picture of the environmental chemistry of the past few centuries, its outlines can be inferred from the limited data

available and are consistent with our knowledge of historical population growth, agricultural development, energy use, and industrial activities.

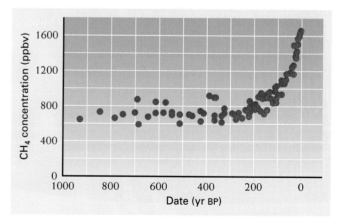

The atmospheric concentrations of CH_4 over the last millenium. The figure shows the time when CH_4 concentrations began to increase rapidly and eventually more than doubled over the last 200 years.

Recent Chemical History of the Urban Atmosphere

The last several decades have been times of accelerated change on Earth. The industrial capacity of the western free market economies and those of the former Soviet Union and other socialist nations increased markedly throughout most of the century. Then, in the 1970s and 1980s, the transformation of undeveloped countries into modern industrial societies began to effect major changes on much of the rest of the Earth. Throughout this period, as we noted above, the human population increased dramatically, especially in the less developed parts of the world.

In the more developed countries, rapid improvements in manufacturing technology and legislation to protect air and water quality brought gradual emission reductions and greater energy efficiency. As a result, emissions into the atmosphere and hydrosphere of trace gases and of particles from these processes have on the whole gradually decreased, *on a unit basis*. For example, the emission of sulfur dioxide per kilogram of combusted coal has decreased at many facilities. On a global basis, however, this technological improvement has been countered by a great increase in the demand for energy, manufactured goods, and food.

The changes in emissions in the past century have recently been documented by Sophia Mylona of the Norwegian Meteorological Institute. Drawing on information pertaining to the use of solid and liquid fuels and the degrees of activity of various industrial processes, she computed levels of sulfur dioxide emissions for all of Europe and much of Western Asia, including Russia, Turkey, and the Baltic states. Mylona estimates that during the period from 1880 to 1991, total sulfur dioxide emissions increased by a factor of about 10 to 12. Emissions peaked in 1975; since that year, vigorous efforts to reduce air emissions have resulted in an overall emissions reduction of 10 to 15% (see the diagrams on page 100).

The revolution that created the modern industrial world coincided with the development of modern analytical chemistry. As a result, it has become possible for changes in the atmosphere's and hydrosphere's composi-

Central Hong Kong at night. The profligate use of energy by a growing world population results in massive emissions from fossil fuel combustion in power plants, degrading the environment on scales from local to global.

tion and chemical reactions to be monitored in detail, even as the transformations are taking place and even in remote locations. Changes in air quality were first noted in urban areas and their immediate environs, where most of the population lived and where most of the measurements were made. In 1845 Christian Schoenbein, a German scientist working at the University of Basle in Switzerland, introduced an "ozonometer," a chemically treated paper that changed color when exposed to ozone. This was extensively used for ozone measurements at many sites throughout the world. Schoenbein's and other's tech-

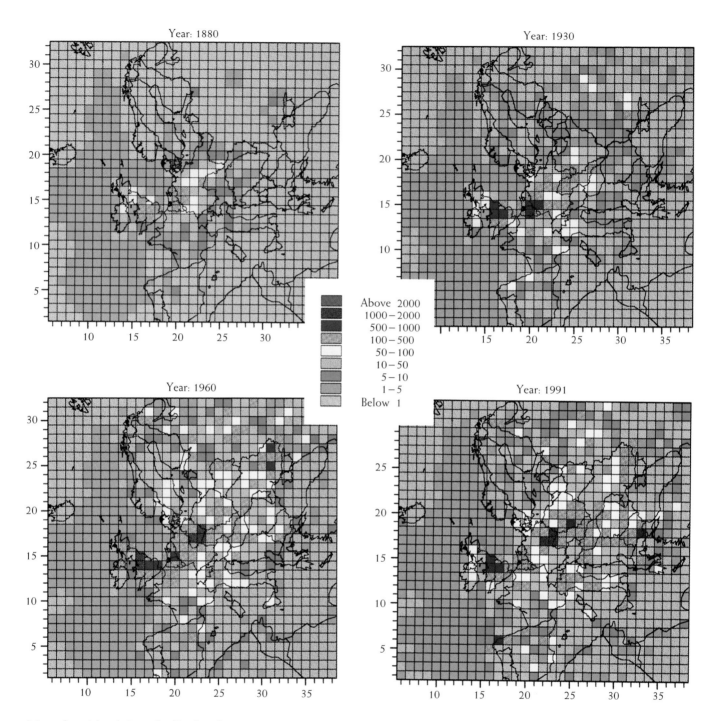

Maps of spatial emissions of sulfur dioxide in Europe and Western Asia for four selected years. Data are expressed in Gg SO₂; to convert to Gg SO, divide by two, the ratio of the molecular weights of S and SO₂.

niques are described on page 102. By the last quarter of the nineteenth century, the ozonometer had clearly demonstrated that ozone was a common constituent of atmospheric air.

By far the most geographically comprehensive assessment of urban air quality is one recently published by the United Nations Environment Program (UNEP). For the period of the study (1980 to 1984), more than half of the participating cities were found to exceed the UNEP guidelines for particulate matter and sulfur dioxide, and these were cities located on nearly every continent. The annual

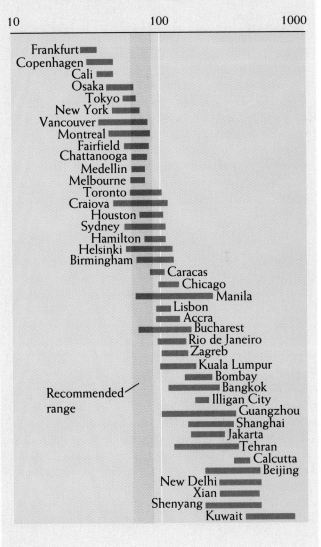

Concentration of particulate matter (µg/m³)

In a scene now common around the world, traffic slows to a crawl in Bangkok, Thailand and the tailpipe emissions inaugurate the chemical reactions leading to urban smog.

The range of annual averages of total particulate matter concentrations measured at multiple sites within cities throughout the world for the period 1980 to 1984. Each bar represents a city, as indicated. Few cities in the developing world have been extensively monitored and therefore do not appear in this figure; many would be expected to have high levels of particulate matter. The orange shading indicates the concentration range recommended by the United Nations Environment Program as a reasonable target for preserving human health.

A Hundred Years of Ozone Data

Schoenbein measured atmospheric ozone by taking starched paper, impregnating it with iodine, and determining the color change corresponding to different ozone concentrations. His measurements in the atmosphere established ozone's presence, but difficulties with standardization and the influence of humidity and wind speed limited the accuracy of the results. To improve on the ozonometer, Albert Levy of the Paris Observatory developed the technique of iodine-catalyzed oxidation of arsenate (AsO_3^{3-}), based on the reaction

$$O_3 + AsO_3^{3-} \xrightarrow{\text{iodine}} O_2 + AsO_4^{3-} \qquad (5.1)$$

Using this technique, Levy measured ozone concentrations daily for 34 years, from 1876 to 1910. If scientists were able to determine that Levy's measurements were accurate, they could usefully compare them with modern measurements and thus discover whether ozone concentrations have changed over time. Fortunately, Levy carefully described what he had done, and his original records are still available in France. Following his procedures, Andreas Volz and Dieter Kley of the German Nuclear Research Center in Jülich constructed a copy of Levy's apparatus in 1987 and exposed it to carefully generated and monitored concentrations of ozone. Their finding was that Levy's experimental approach did indeed give reliable ozone concentration values if slight corrections were introduced to take into account a sensitivity to sulfur dioxide in the air. When these adjustments were made, Levy's data revealed concentrations of ozone averaging 10 ppbv. Present-day average annual ozone concentrations at low-altitude stations in rural Germany, which one might expect to be reasonably similar in air quality to a Paris suburb of the nineteenth century, are actually more likely to be in the range of 25 to 35 ppbv, with average summer (April to September) values of 35 to 45 ppbv. Occasionally during summer smog episodes in those areas, maximum ozone values may reach 100 ppbv, values that were never reported in the old data sets. Clearly, there have been enormous increases in background ozone concentrations over the past half century, especially downwind of heavily developed urban and industrial areas.

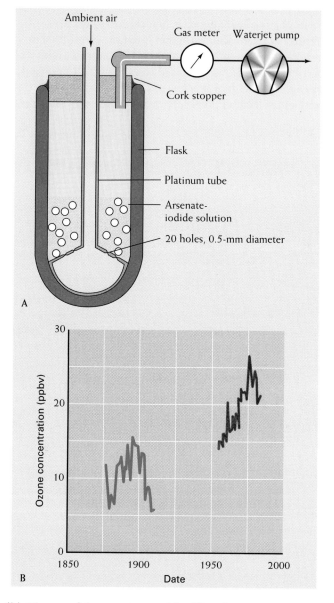

(A) Diagram of the apparatus used by Albert Levy to measure ozone concentrations at Montsouris (near Paris), France, from 1876 to 1910. (B) A comaparison of Levy's surface ozone concentrations (blue) with those measured at Arkona (a German island in the Baltic Sea) from 1956 to 1982 (purple).

mean levels of atmospheric particles ranged from a low of about 35 micrograms per cubic meter ($\mu g\ m^{-3}$) to a high of about 800 $\mu g\ m^{-3}$, a range of a factor of about 25.

UNEP global data for urban sulfur dioxide shows that in more than a quarter of the cities studied, the annual average sulfur dioxide concentrations are at levels thought to have a high potential for causing severe corrosion impact. The difference between the highest and lowest average concentrations of sulfur dioxide in the 54 cities surveyed is nearly a factor of 100. The half-dozen cities with the highest annual averages include locations in Europe, Asia, and South America.

UNEP also studied *trends* in air quality data. Of the 37 cities surveyed from 1980 to 1984, six exhibited strong upward trends in the concentration of atmospheric particulate matter. These included urban areas in South America, Asia, and Europe, where air quality is rapidly worsening as industrialization progresses while emissions control does not. Many less developed countries do not yet have active air monitoring programs, but it seems likely that at present most of them have increasing levels of emissions and, consequently, a decreasing quality of air.

More detailed information concerning emissions trends is available from a number of countries that have long conducted atmospheric monitoring programs, and these data have been used to demonstrate the effects of regulatory actions taken to improve air quality. Perhaps the most dramatic evidence for improvement in an air quality component as a result of legislation is that for airborne lead in the urban areas of the United States, shown in the diagrams on the following page. A 12-year study of ambient lead concentrations from 1971 to 1983 shows that the substantial reductions in leaded gasoline use in the United States during that period were reflected in sharply decreasing atmospheric lead concentrations. As of 1983, the mean lead level had decreased from about 1.0 $\mu g\ m^{-3}$ to about 0.15 $\mu g\ m^{-3}$. Air traveling from the United States is an important component of the atmospheric gases trapped in Greenland ice cores, and lead data from those cores show strong decreases as well.

4170 meters above sea level on Mauna Loa, Hawaii, an atmospheric chemist collects samples of air wafted halfway across the Pacific Ocean. Samples acquired over many years from sites such as this document changes in the composition of many of the atmosphere's constituents.

Changing Chemistry

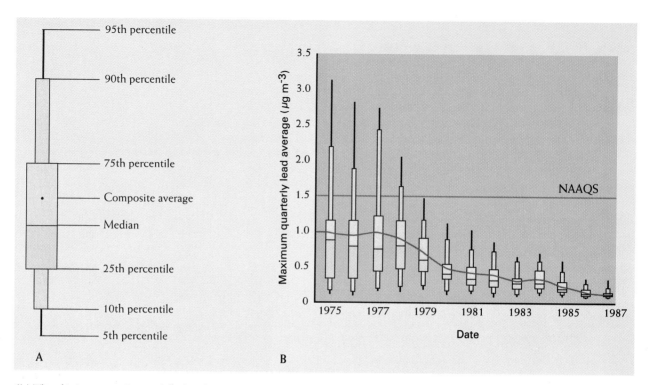

(A) The plotting convention used for boxplots. (B) Boxplot comparisons of trends in maximum 3-month average lead levels at sites in the United States from 1975 to 1987. NAAQS is the U.S. National Ambient Air Quality Standard for atmospheric lead.

Recent Chemical History of the Global Atmosphere

Global patterns of chemical change are more difficult to pinpoint than urban changes are, in part because much of the data must be collected from places where the air is unperturbed by local emissions and thus more sophisticated measurement techniques are required. Carbon dioxide was the first atmospheric gas to be routinely monitored with high precision at a site reflecting the global background situation, in a remarkable effort begun in 1958 at Mauna Loa, Hawaii, by David Keeling of the Scripps Institution of Oceanography. The seasonal cycle of carbon dioxide fluctuations produced by photosynthesis and respiration of vegetation at temperate and subtropical lati-

tudes was clearly evident in the data, as was the strong long-term upward trend. In a similar vein, but over a somewhat shorter time period, data collected by the state of New Jersey from 1968 to 1977 and analyzed in 1980 by Thomas Graedel and Jean McRae of Bell Laboratories suggested an increase in atmospheric concentrations of both carbon monoxide and methane. Since that time, extensive and carefully calibrated measurements from all over the world have confirmed those findings. Moreover, the concentrations of methane as measured in ice cores for the years prior to about 1965 blend smoothly into the modern gas chromatographic measurements of the past two decades. The combined data show a steady increase over the entire period. We have pointed out that methane has many sources, most of them related to human activity: coal mining, leaks in natural gas distribution systems, rice cropping, and fermentation in the guts of ruminant animals, especially cattle. Emissions from all these sources have increased substantially over the past 200 to 300 years.

More immediately relevant to the condition of the biosphere is the recent history of concentrations of stratospheric ozone, the main atmospheric absorber of ultraviolet radiation. Routine measurements began in the 1930s, taking advantage of the fact that ozone, like other molecules, absorbs radiation at specific wavelengths. The workhorse measuring instrument that eventually came to be utilized at more than three dozen sites on Earth's surface was developed by George Dobson of England; it consisted of a photometer that monitored light from the Sun within several adjacent wavelength bands. Changes in the relative amount of light received in the various bands have been shown to be related to the total amount of ozone between the instrument and the Sun—that is, from the bottom of the atmosphere to the top. Since measurements are made at different times of day, looking through the atmosphere along different path lengths and slant angles, some idea of ozone concentration as a function of altitude can be derived as well. When we compare the historical data with the results of modern chemical and optical measurements taken in similar locations, it is clear that tropospheric ozone levels are increasing and stratospheric ozone levels are decreasing.

The value of a long-term record for a crucial atmospheric species is indisputable, yet the slow pace at which such information accumulates has often blunted bureaucratic enthusiasm for atmospheric monitoring efforts and made it difficult for scientists to procure adequate institutional and financial support. However, a hypothesis proposed in 1974 by Mario Molina and Sherwood Rowland, both then of the University of California at Irvine, has provided a strong incentive for continued ozone monitoring; they proposed that man-made chlorofluorocarbons (CFCs) could seriously perturb the atmospheric ozone layer. Molina and Rowland theorized that CFCs, especially the chemically very stable refrigerant and propellant gases CFC-11 ($CFCl_3$) and CFC-12 (CF_2Cl_2), are not broken down in the troposphere by reactions with the hydroxyl radical or any other potential oxidizing constituent. Instead, over a period of a few years these gases move from their release points at Earth's surface up into the stratosphere. Above about 20 to 25 km the available solar radiation is energetic enough to destroy them, thereby releasing chlorine atoms ($Cl\cdot$) and chlorine

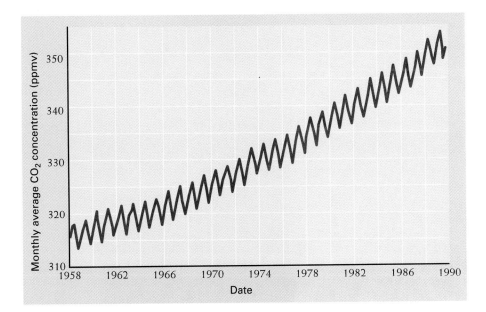

Concentration of atmospheric carbon dioxide at Mauna Loa Observatory, Hawaii, 19.5°N, 155.6°W, for the period 1958 to 1988.

monoxide radicals (ClO·) that are powerful ozone-destroying catalysts. The relevant overall reaction (illustrated for CFCl₃ and actually involving several steps) is

$$CFCl_3 + h\nu \text{ (wavelength } < 260 \text{ nm)} \rightarrow$$

$$CO_2 + HF + 3Cl· \text{ or } 3ClO· \qquad (5.2)$$

followed by the catalytic ozone destruction cycle discussed in Chapter 3. Minor amounts of chlorine are also supplied to the stratosphere by occasional volcanic eruptions and by the upward transport of methyl chloride (CH_3Cl), a product of seaweed and biomass burning. The natural chlorine content of the stratosphere from these latter sources is about 0.6 ppbv. The quantity supplied by CFCs and other industrial organic chlorine compounds is currently about 3 ppbv, some five times the natural background.

The ozone decline due to CFC chemistry was predicted to be significant, but slow. Thus, the world was shocked in 1985, when Joseph Farman and his colleagues at the British Antarctic Survey reported that beginning in the mid-1970s, there had been extremely sharp decreases in the atmospheric ozone concentration over Antarctica every September and October. This is the time of the Austral spring, when 24 hours of darkness gives way to 24 hours of daylight. Data complementary to the British ground-based ozone column records were then provided by a United States orbiting satellite, that had been measuring the ozone column from above for over a decade (and over most of the globe rather than at a single point). Because glitches in satellite data transmission occasionally result in anomalous zero readings that could distort computer-averaged values, the NASA computers had been programmed to ignore ozone concentration values thought too low to be realistic. Once this artificial impediment was removed, the ozone hole's existence was revealed by the satellite data as well. Two important points emerged from the combination of ground-based and space-based observations: the region of decreased ozone extends over the entire Antarctic continent and beyond and, with minor variations, less and less ozone is present with each passing year.

Within only a few years this alarming rate of decrease was traced to the presence in the Antarctic stratosphere of frozen particles that consisted of nitric acid (HNO_3)

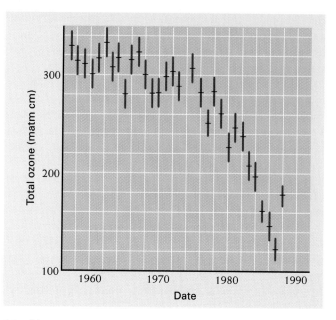

Monthly mean values (horizontal lines) and ranges (vertical bars) of total ozone over Halley Bay, Antarctica, as measured by the British Antarcic Survey and NASA's total ozone monitoring experiment for October of the years 1957 through 1993. A layer of gaseous ozone 1 mm thick at 1 atmosphere pressure and 0°C corresponds to 100 matm cm. The original version of the figure, published in 1985, is one of the two most famous graphs in atmospheric chemistry; the Keeling CO_2 curve shown on page 105 is the other.

and water molecules and served as catalytic sites for a series of chemical reactions. As you will recall from Chapter 3, the chlorine atom is a very efficient destroyer of ozone, but its destructive nature can be temporarily suppressed when it is incorporated into a reservoir molecule such as chlorine nitrate ($ClONO_2$) or hydrochloric acid (HCl). In principle, these chlorine-containing reservoir species can react with each other as follows:

$$ClONO_2 + HCl \rightarrow Cl_2 + HNO_3 \qquad (5.3)$$

with chlorine radicals being subsequently produced when the Cl_2 molecules are photodissociated in sunlight of wavelengths shorter than 450 nm:

$$Cl_2 + h\nu \rightarrow 2Cl· \qquad (5.4)$$

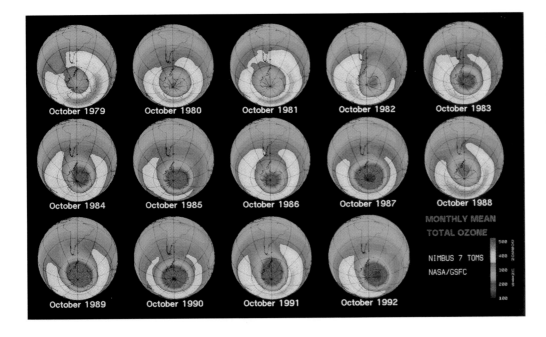

Total Ozone Mapping Spectrometer (TOMS) satellite observations graphically depict the decline of atmospheric ozone over the South Pole between October 1979 and October 1990. The amount of ozone present is indicated by color: green represents an average amount of ozone, blue less, and purple still less. From 1979 to 1990, ozone concentrations fell by more than 50%.

Reaction 5.3 occurs so slowly in the gas phase, however, that it is inconsequential. But when frozen particles of nitric acid ice or water ice are present, both $ClONO_2$ and HCl readily attach to the ice surface and react with each other, and in that medium reaction 5.3 proceeds much more quickly than in the presence of gas alone. Once the products of the reaction have formed, Cl_2 vaporizes into the surrounding air, whereas the HNO_3 is retained in the ice matrix. The reaction may thus be written

$$ClONO_2 + HCl \xrightarrow{ice}$$

$$Cl_2(gas) + HNO_3(ice) \quad (5.5)$$

As a consequence of reaction 5.5, the nitrogen dioxide that might otherwise be regenerated from the chlorine nitrate and interfere with the cycle of ozone loss is not available to do so. Therefore, the ozone remains unprotected against destruction by the powerful Cl· and ClO· radicals.

In its polar location, Antarctica receives no solar radiation for several months each year and stratospheric temperatures cool to −80°C or colder, sufficient to form the nitric acid-ice or pure ice particles that facilitate re-

"What would you recommend for the ozone hole?"

Enhanced ultraviolet radiation from the sun increases the potential for skin cancer. Fashion rises to the challenge in this approach of the millinery industry to ozone hole concerns. (© 1992, The New Yorker Magazine, Inc.)

Effects of Ultraviolet Radiation
on Biological Organisms

*T*he changes in the concentration of stratospheric ozone are expected to have a direct impact on biological systems. Currently, living things are protected from excess ultraviolet radiation because the penetration of solar radiation to Earth's surface at wavelengths longer than about 220 nm is strongly limited by ozone absorption. This absorption extends to wavelengths just over 300 nm, although with rapidly decreasing efficiency. As a result, the amount of 290-nm radiation reaching the surface is less than the amount of 320-nm radiation by a factor of about 10,000.

Solar radiation provides heat, light, and energy to plants and animals, and modern ecosystems have adapted to utilize these resources to the fullest extent possible. Thus, biota tend to thrive under conditions similar to those under which they evolved—and to deteriorate when those conditions change significantly. The tolerance of organisms to photons in the wavelength band between 290 and 320 nm, termed the UV-B (ultraviolet B) region, exemplifies this adaptation. These are wavelengths which, if absorbed, damage DNA molecules, and the shorter of these wavelengths (corresponding to the photons with the greatest energies) are those to which organisms are more sensitive by a very large margin. Any significant decrease in ozone concentrations causes large increases in ground-level UV-B and, in turn, in the DNA damage that occurs. This situation is a product of evolution: organisms with DNA susceptible to the green or red radiation, the radiation that *always* penetrated to the planetary surface, would not have survived over the long term. This interrelationship between stratospheric ozone and DNA, the central molecule of life, renders maintenance of the stratospheric ozone shield a high priority for the continuance of the biosphere as we have come to know it.

When a wavelength-dependent comparison is made between the flux of solar radiation at Earth's surface (that is, the radiant energy per unit area) and the sensitivity of biological organisms to the radiation, the elegant operation of the DNA molecule is once more apparent. The top graph shows how rapidly the sensitivity changes just at the wavelengths protected by stratospheric ozone. Increased radiation fluxes at those wavelengths have detrimental effects for all organisms, though the degree and type of effect differ for different species. John Fredericks and his colleagues at the University of Chicago have studied the wavelengths and strengths of solar radiation reaching the surface of the Antarctic continent. Such research demonstrates that the lower ozone concentrations in the Antarctic stratosphere are indeed allowing increasing levels of DNA-active radiation to penetrate to the Antarctic surface.

Humans The DNA in the cells of human skin is susceptible to damage by UV-B radiation. Particularly for light-skinned people, the most sensitive, the result can be cancerous growth of the squamous cells (the skin's outer layer), the basal cells (the intermediate layer), or the melanocytes (the deepest layer). Studies have shown that a 1% reduction in the ozone layer increases the effective UV-B dose at sea level by about 2%. This increase in turn leads to about a 4% increase in the incidence of basal cell carcinoma and about a 6% increase in squamous cell carcinoma. With a 10% reduction in stratospheric ozone, the latter cancer incidence rates increase about 50% and 90%, respectively. Thus in Germany, for example, a 10% reduction in stratospheric ozone would lead to approximately 20,000 additional cases of skin cancer each year. The probability of occurrence of all skin cancers is sharply reduced by screening out UV-B radiation. Other effects of increased UV-B radiation on humans include a higher incidence of cataracts and a weakening of the immune system.

Reptiles Some reptiles lay their eggs in concealed locations or underwater; others leave them boldly exposed on dry land. The recent increases in UV-B radiation at midlatitudes seem to be reducing the proportion of exposed eggs that produce healthy young. Not only are these species threatened with eventual extinction, but their predators will lose a local food source, causing disruption of the entire local ecosystem.

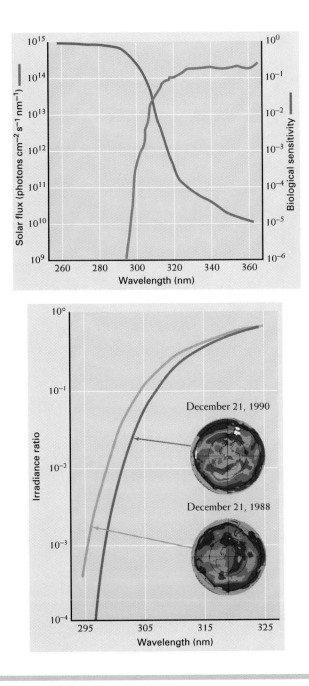

Plants   Approximately 200 plant species have thus far been examined for sensitivity to UV-B radiation. About 50% demonstrate significant adverse effects, including reduced average leaf area, reduced shoot length, and decreases in the rate of photosynthesis. The most complete information has been gathered for agricultural crops; it suggests, among other findings, that a 25% reduction in stratospheric ozone would lead to a 50% reduction in soybean yield. There is also preliminary evidence that natural nitrogen fixation may be inhibited by increased exposure to UV-B.

Marine ecosystems   Even under normal conditions, many species of plankton are sensitive to low-level doses of UV-B. The defense mechanism they have developed over evolutionary time is to live at a water depth that screens out most of the UV-B but allows the longer wavelength radiation used for photosynthesis to reach them. A loss in protective atmospheric ozone would thus imply either direct harm from the increased radiation or indirect damage from a decrease in photosynthetic activity as the plankton retreat to greater depths. Because plankton occupy a position near the bottom of the food chain, a reduction in their numbers would have negative consequences on the higher-level members of the ecosystem, such as zooplankton and fish.

(Top) The solar flux reaching Earth's surface when the Sun is at a zenith angle of 39° (red line) and the absorption by DNA of solar radiation, ratio to absorption at 260 nm (purple line). (Bottom) Ground-level irradiance ratio (the ratio of received solar flux at the indicated wavelength to that at 350 nm, where no ozone absorption occurs) at Palmer Station, Antarctica. The data are for local noon on December 21, the austral summer solstice, in 1988 and 1990. The Antarctic ozone hole in 1988 was deep, but not catastrophically so, and had nearly closed by December 21 of that year. In 1990, however, the ozone hole was deeper and more persistent, and on December 21 was still present. The relative impact was biologically active ultraviolet radiation was substantial, with noon irradiances at 298 nm about 10 times larger in 1990 than in 1988.

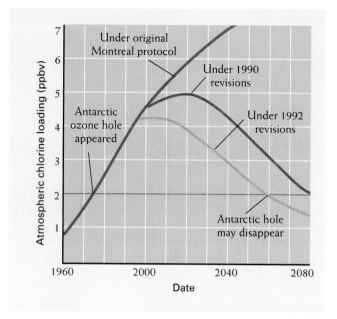

Chlorine abundances in the stratosphere, measured (1960 to 1992) and projected (from 1992 on under the terms of the Montreal Protocol on Substances That Deplete the Ozone Layer (1987), the London amendments (1990), and the Copenhagen amendments (1992).
The Antarctic ozone hole appeared in 1975 when the chlorine loading reached about 2 ppbv; it is anticipated that under the most restrictive of the regulatory plans it will not disappear until the chlorine loading decreases below 2ppbv in about 2060.

action 5.5. As the Sun makes its appearance in late September, the chlorine photodissociation of reaction 5.4 begins and ozone depletion occurs. The destruction is exacerbated by the atmospheric circulation pattern over Antarctica in winter, which is dominated by an almost circumpolar airflow, the polar vortex, that holds the chlorine oxide–rich air over Antarctica for many months.

Not all the news about the changing atmosphere is bad. Indeed, some global air-quality indicators have recently begun to change for the better. The shining example is the marked decrease in the growth of the concentrations of the two most widely used chlorofluorocarbons, CFC-11 and CFC-12, which constitute most of the stratospheric burden of chlorine. Because of concerns about the ozone depletion by these and other CFCs, an international agreement, the Montreal Protocol, was signed in 1987, binding the signatories to limit the production of

these compounds. Most of the principal CFC-producing countries participated, as did many of the less highly industrialized nations. The intent of the protocol was to lower the rate of increase of these CFC molecules in the stratosphere, and eventually (but not for many decades) to cause their concentrations to decrease to low levels, as shown in the illustration here. That agreement was later strengthened by amendments at meetings in London in 1990 and Helsinki in 1992, largely by hastening the transition to hydrogen-containing CFCs (HCFCs), which react with HO and are thus largely removed in the troposphere. The HCFCs still contribute to stratospheric chlorine, but at much reduced rates. Atmospheric CFC-concentration data have been obtained through increasingly precise and extensive measurement networks during the past decade. Those data, shown in the graphs below, indicate that the mixing ratios of CFC-11 and CFC-12 in the lower atmosphere, which have been climbing

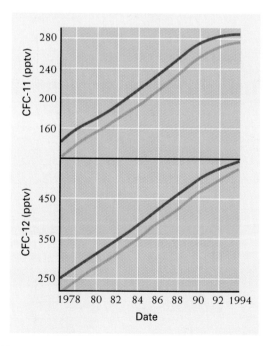

Monthly means of observed mixing ratios of CFC-11 and CFC-12 at seven sites around the world for the interval 1978 to 1993. Calculated mean concentrations are shown for the northern hemisphere (purple curve) and southern hemisphere (orange curve).

since their invention and deployment in the 1930s, began to level off in 1990. Should the further anticipated decreases in production be realized, it is estimated that CFC-11 and CFC-12 mixing ratios will reach a maximum in the troposphere within the next few years. Maximum levels in the stratosphere will be reached 5–10 years later because of slow rates of troposphere–stratosphere exchange. The stratospheric levels will then begin a slow decline. Since CFC-11 and CFC-12 are the most abundant atmospheric CFCs, this would confirm the projections of the graph on the facing page. As with other pollutants that have been subjected to stringent emissions controls, the CFC-11 and CFC-12 story demonstrates that when serious problems are identified, intervention can indeed produce results visible on short to moderate time scales.

But even though the *rate* of increase of the principal CFCs is decreasing, the *concentrations* of CFCs are still increasing as more are added to the atmosphere year after year. A similar situation holds for a number of other trace gases with anthropogenic sources. The table on this page gives some statistics for a few species whose concentrations have been carefully studied at a number of sites around the world for a decade or more: average concentrations, rates of change, percentage rates of change. The increases listed here show that the global budgets of these gases are out of balance, that is, the inputs and outputs are not equal. Budget imbalance is not an issue in regional budgets, because spatial variability is always present and some regional budgets may be out of balance to the extent and direction needed to balance other regional budgets. Also, some budget elements are so difficult to quantify that it may sometimes be hard to tell whether a budget is in balance or not. When sufficient information is available, however, a well-validated but out-of-balance global-scale budget indicates a definite global change. The central message of the table is that for atmospheric species emitted as a consequence of human actions, most budgets that have been assessed over the past few years are out of balance in the additive direction, that is, the inputs outweigh the outputs. Unless these budgets are promptly brought into balance, the planet will become more and more contaminated with the passage of time.

We clearly see changes in the atmosphere's chemistry over times long and short, and at altitudes high and low. The carbon dioxide and methane data, extending nearly 200,000 years into the past, show highly suggestive correlations with temperature. Do they become more abundant when global temperature is high, or does their high abundance *cause* global temperature to be high? We are not yet certain. Much more certain are the changes that have been seen over the past few decades in our cities—those changes can be directly traced to human activities. Finally, on the global scale and within our lifetimes, the budgets of many anthropogenically produced gases have gone out of balance (in particular, carbon dioxide concentrations have increased rapidly), and stratospheric ozone (in some places and at some times of year) has fallen abruptly. What these trends may mean for the future will be explored in the next two chapters, as we survey the contributions of computer modeling research.

The Assessment of Imbalance in Budgets of Long-Lived Atmospheric Species

Species	1990 global average concentration	Estimated 1990 annual concentration change	Estimated 1990 annual percentage change
Carbon dioxide (CO_2)	354 ppmv	+ 1.8 ppmv	+ 0.5
Methane (CH_4)	1.72 ppmv	+ 10.0 ppbv	+ 0.6
Nitrous oxide (N_2O)	310 ppbv	+ 0.8 ppbv	+ 0.25
Molecular oxygen (O_2)	20.9%	− 0.4 ppmv	− 0.002

The Bay of Bengal in Bangladesh, a country of high population and low altitude. Often ravaged by tropical storms, Bangladesh is particularly susceptible to any change in sea level that could accompany a global warming.

Predicting the Near Future | 6

You take my house,
when you do take the prop
That doth sustain my house;
You take my life,
When you do take the means whereby I live.
—William Shakespeare, The Merchant of Venice

*H*umans are inherently conscious of and curious about the future. We know that changes can and will occur—in our government, in our economic situation, and in our family responsibilities—and we try to plan our lives with those prospects of change in mind. In the last decade or two, another concern has been added to our list: the possible consequences of changes in the Earth system. The natural questions to ask then are, "Given what we know about changes that have occurred in the distant and more recent past, can we predict the characteristics of tomorrow's atmosphere and climate? Or those of the next century? Or the next millennium?" Atmospheric scientists can indeed, with reasonable confidence, forecast the patterns of the next few years, but their predictions become increasingly uncertain as they attempt to describe the more distant future. This chapter introduces the tools scientists are using to make those predictions—and considers what the results of this research may mean for our future.

Models of the Earth System

It is becoming more and more commonplace in the environmental sciences to describe complex systems of interacting physical, chemical, and biological processes through the design of numerical "models." These models consist of sets of mathematical equations that attempt to describe processes seen in nature, allowing scientists to create replicas of natural systems with a computer so that the causes and effects of system behavior may be better understood. Although we can study individual interactions within a system by using laboratory simulations or, under favorable conditions, by directly observing nature, the complexity of Earth system processes makes the use of these mathematical models necessary as well, in order to comprehend the behavior of the system as a whole.

From a scientific point of view, the most interesting stage of model development is when knowledge has advanced to the point that researchers can design a reasonable model but is not so complete that all variables influencing the system are well understood. At this stage, the model's results often deviate from actual measurements, leading scientists to search for contributing processes that they have not yet considered. The chemical models of the Earth system are at this intermediate, innovative stage right now. As a result, the calculations being performed today are likely to improve our understanding of atmospheric chemistry markedly and unpredictably. In just such a way it was discovered that photochemical processes in the gas phase alone could not explain the rapid loss of ozone under ozone hole conditions, a finding that led to the discovery of several important reactions that take place on ice particles. Atmospheric ice particles do not exist only within the ozone hole, so this finding stimulates research progress pertaining to other regions of the atmosphere as well.

Once scientists are sufficiently convinced of a model's validity, they begin to use the model to predict future conditions arising from changes in important variables within the system: natural processes, such as solar activity or volcanic eruptions, or anthropogenic factors, such as emissions of industrially produced trace gases. Models can also be used to study the past, which in turn provides additional opportunities for testing the models or for learning more about the effects of such things as continental drift or bolide impacts. It is this ability of models to explore situations remote from

Marine scientists launch an instrument to measure the velocity of undersea ocean currents predicted by computer models.

current conditions and unavailable in reality—to conduct numerical experiments that ask, "What would happen if . . . ?"— that makes them so potentially valuable.

A current example of the significant role that numerical models can play is their use to simulate the changes that might result from different regulatory or political actions, such as the phase-out of chlorofluorocarbons, or from a change in the relative proportions within the current mix of modifiable energy sources: coal, oil, natural gas, and nuclear. In these applications, models become policy tools, and much responsibility accrues to the scientific community to carefully test the models and to communicate the necessary caveats along with the results. Because of our policymakers' urgent need for advice, some atmospheric chemical models are already used in this way, although the models are not yet (and probably never will be) considered fully reliable. Notwithstanding these reservations, current models represent some of the best tools available for assembling large amounts of information into a form that can be critically appraised and used.

Numerical atmospheric models differ greatly in complexity, especially in their degree of spatial and temporal resolution. Ideally, a model should be able to generate predictions specific to large and small geographical areas and for long and short time intervals. In practice, though, the information required to do that may not be available, or the capabilities of the computers may be insufficient. All computer modeling efforts at best represent tradeoffs: scientists must choose spatial and temporal resolution at the expense of physical, chemical, and meteorological detail, or vice versa. For example, smog chemistry may be calculated in great detail, but only for a single city, or climate could be calculated for a century, but only for continents instead of individual regions. Each choice has its uses and limitations; no one model can serve all needs.

Earth system models have several characteristic formats. The simplest model for atmospheric applications is the box model, envisioned as a box into which some things are added, from which some things are taken away, and within which changes occur, see the diagram on the following page. Chemical species enter in two ways: they are emitted from sources within the box, or they enter by entrainment (the addition of air and its chemicals into the box from the surroundings as a consequence of atmospheric motions). Conversely, detrainment (the loss of air and its chemicals into the surroundings) represents a loss of chemical species. Often a box model of atmospheric chemistry is "placed" at ground level at a specified location so that the chemical effects of emissions from the geographical region into the air immediately above it can be studied.

The dimensions and placement of the box are normally dictated by the particular problem of interest. If one wishes to study the influence of urban emissions on the chemical composition of the air leaving an urban area, for instance, then the box may be designed to cover the entire urban area, and emissions may be assumed to be mixed thoroughly throughout the box. Chemical species from outside the box are "blown" into the box. The chemical composition of the air within the box is then determined by the model. Chemical reactions must, of course, be taken into account. Because a few of the reactions are driven by solar radiation, appropriate wavelength-dependent solar fluxes must be incorporated into the calculations. The reactions will be sources of new chemical species as well as sinks for existing molecules and atoms. Not only gas-phase reactions but also chemical reactions on the surfaces of aerosol particles, in cloud or fog droplets, or in raindrops are often considered. Many investigators are at least as interested in the *products* of the chemical reactions as in the primary emissions themselves, such as the generation of ozone from hydrocarbons and oxides of nitrogen and the formation of sulfate particles from sulfur dioxide.

In many calculations in traditional scientific fields, the relevant equations are set up, the required input data supplied, and, if everything is formulated properly, the computer does its work and the result appears. An Earth system calculation, however, is generally more complicated because of the investigator's interest in the system's progress over time. In other words, most models of atmospheric chemistry are time dependent. Thus, the results of the first calculation must be supplied to a new calculation, along with appropriate new information on emissions, radiation, and so forth, in order to compute a result for a later stage in the system's development. The interval between the times represented by the sequential calculations is known as the "time step"; it can be large if all the factors are changing slowly, but must be quite small if factors (such as emissions from rush-hour traffic) are rapidly changing. A complete model result, therefore, may encompass hundreds or thousands of calculations as scientists attempt to mimic the changing world from the beginning to the end of the time span of interest.

An example of a time-dependent box model might be one designed to compute urban air quality for a 24-hour day. Any variations over that period of time in any of the influential factors would need to be accurately supplied to the model; these variables would certainly include hourly variations in traffic and other pollution sources, diurnal

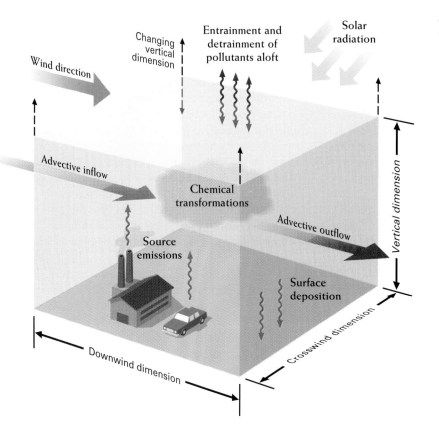

Wind direction

Changing vertical dimension

Entrainment and detrainment of pollutants aloft

Solar radiation

Advective inflow

Chemical transformations

Advective outflow

Source emissions

Surface deposition

Vertical dimension

Crosswind dimension

Downwind dimension

wind-speed patterns, variations in the height of the well-mixed layer in the lowest kilometer or so of the troposphere, and the variation of solar radiation.

To construct a box model, first the builder defines the dimensions of the box and an appropriate approach to time variations. Next, pertinent variables—source fluxes, surface depositions, advection and entrainment rates, and so forth—are represented by mathematical formulas, and initial and boundary flux conditions are specified. The suitable description of each of these factors has been and continues to be a topic of inquiry, so representing them mathematically is a complicated exercise requiring many decisions, some well constrained, some not. The appropriate chemical reactions must then be decided, also potentially difficult because Earth systems are chemically so complex that often not all reactions of interest are identified. Furthermore, sometimes the rate constants or absorption coefficients for a known reaction have not been determined in the laboratory, in which case estimates must be used instead. There is always the risk that some key processes have not even been discovered yet or that chemical reactions that in the past were rather unimportant have, unbeknownst to the investigators, become important following human-induced changes in atmospheric conditions.

The box model is sometimes called a "zero-dimensional" model, because of its inability to calculate variations in any spatial dimension, horizontal or vertical. Rather, the set of results it computes at each time step is an estimate of the average meteorological conditions and chemical composition within the box at a point in time. As a consequence, box models are generally best used for taking a first look at a particular problem. They are not intended to be definitive but to provide a quick and rough idea of the most important processes, without the tedious and exacting requirements for determining the details of horizontal wind speeds, vertical mixing times, emissions fluxes in different locations, and the like.

The simplest improvement over the zero-dimensional box model is to compute changes in the concentrations of chemical species as a function of a single spatial dimension—an approach that allows the model to capture, for example, the variation of ozone with height. The earliest models of this type were constructed for studying the chemistry of the middle atmosphere (the stratosphere and mesosphere), and therefore altitude was chosen as the variable dimension. In more recent years the troposphere has also been simulated with one-dimensional models, again with altitude as the variable dimension.

One-dimensional (1D) models can be visualized as a series of box models, one on top of the other. The layers can be designated as equal in depth, but in practice most 1D models are designed to have more and thinner layers near the ground, where interactions with the surface are extensive and meteorological exchange processes more variable, and fewer and thicker layers farther from the planetary surface, where chemical and dynamical variations with altitude are less. A typical 1D model might use 10 or 15 layers and must deal with the influence of each layer on its neighbors.

The builder of a 1D model confronts several tasks in addition to those required of the box model builder. Initial conditions must now be defined for each of the layers, and transport between the layers must be represented

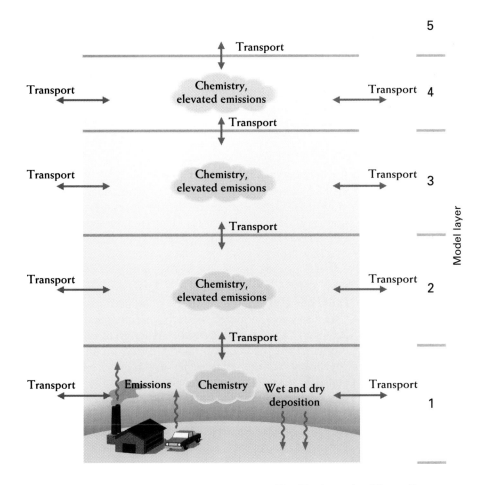

A schematic diagram showing the features of a 1D model of atmospheric chemistry. The elevated emissions entering other than at the ground come from tall exhaust stacks, aircraft, or volcanoes, or surface emissions rapidly drawn upward in convective clouds.

appropriately. The characteristics (chemical composition, humidity, and so on) of the air entering each layer by advection must be specified. The calculation then proceeds as with the box model, time step by time step, but with an additional complication: after each time step the results from the previous one must be used not only to start the next, but also to adjust the input and output factors in the adjacent atmospheric layers. Through this process, for example, emissions at the ground (in layer 1) rise gradually through the layers, evolving chemically on the way.

The Effects of Chlorofluorocarbons on the Stratosphere

A typical application of 1D models, and one that continues to be very useful for scenarios in which the chemistry is so detailed there is little room left for meteorology, is the troposphere–stratosphere model. In this application, each of the vertically stacked layers represents a global average at a given altitude. By definition, the lateral transport of chemical compounds is ignored, and the interaction between layers is described by the exchange of air parcels.

A typical example of a 1D model for the Earth's troposphere and stratosphere is one developed by Christoph Brühl and Paul J. Crutzen, of the Max Planck Institute for Chemistry in Germany, to study the chemical reactions thought to be responsible for the depletion of stratospheric ozone. In their study, Brühl and Crutzen investigated the effects on stratospheric ozone of a number of different emission scenarios. Each scenario assumed the same pattern of concentration with time of the following species: carbon dioxide, methane, nitrous oxide, carbon monoxide, and NO_x (NO + NO_2). Chlorofluorocarbon (CFC) emissions are treated separately, in several scenarios. One

(Scenario I) assumed continued CFC production at 1974 levels. Another (Scenario II) assumed the CFC production restrictions specified in the Montreal Protocol of 1987: a 50% reduction for both CFC-11 and CFC-12 by 1998. A third (Scenario III) assumed the CFC production restrictions specified in the 1990 London amendments to the Montreal Protocol: for CFC-11 and CFC-12, maximum allowed percentages of 1986 production are 80 (1993), 50 (1995), 15 (1997) and 0 (2000); in addition, the cessation of methyl chloroform and carbon tetrachloride production by 2005.

The Scenario I results summarized in the top graph on the right demonstrate that if emissions of CFCs were to continue indefinitely at 1974 rates, the ozone concentrations at 40 km altitude could be reduced by as much as 30% by the year 2005 and by nearly 50% by the year 2030. If the Montreal Protocol were fully implemented (Scenario II), it would have little effect by the year 2005, because the reductions would just have begun; however, less additional degradation would be anticipated between the years 2005 and 2030. Were the London amendments to be followed (Scenario III), little effect would again be seen by 2005, but by the year 2030 significant improvement would be evident. In Brühl and Crutzen's study, it is significant that the results suggest very substantial negative impacts on column ozone early in the next century if CFC emission reductions do not occur as scheduled.

CFCs impose a dual burden on the atmosphere. Not only do they contribute to stratospheric ozone depletion, but they also absorb infrared radiation and thus enhance the greenhouse effect and consequent global warming. The likely replacement substances for CFCs, at least in the short term, are partially hydrogenated variants termed HCFCs (hydrochlorofluorocarbons). Because the added hydrogen atoms permit these compounds to react with the hydroxyl radical, HCFC molecules are shorter lived and much less likely to reach the stratosphere than are CFC molecules.

The realization of the potential effects of CFCs and HCFCs, and the implications of that realization for public policy, have created a need for relatively simple ways of comparing the impacts of the different gases. In the case of ozone, this comparison is accomplished with 1D numerical models that compute ozone depletion potentials (ODPs) for the gases. The ODP represents the amount

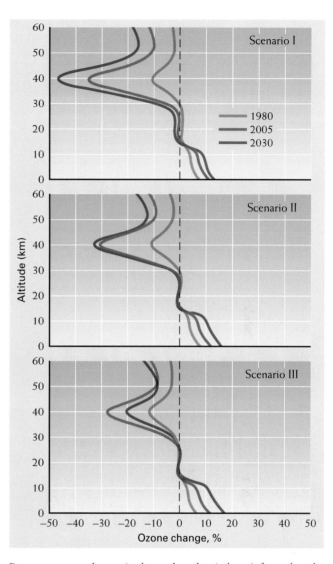

Percentage ozone changes in the southern hemisphere (referenced to the levels calculated for the year 1965), computed with a 1½D model (consisting of two related 1D models, one for each hemisphere) for three different chlorofluorocarbon emission scenarios. Scenario I assumed constant CFC emissions at 1974 levels. Scenario II assumed CFC emission reductions as specified in the Montreal Protocol of 1987. Scenario III assumed CFC emission reductions as specified in the 1990 London revisions to the Montreal Protocol. While the results provide a graphic demonstration of the capabilities of 1D models, this work was completed before the most recent advances in CFC chemical understanding, so it is now thought that these calculations somewhat underestimate the actual ozone depletion.

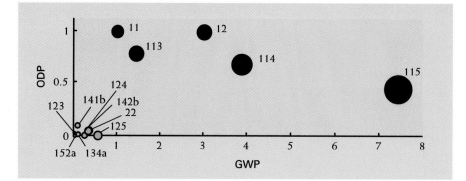

Relative global warming potentials (GWPs) and ozone depletion potentials (ODPs) of several chlorofluorocarbons (CFCs; black circles) and hydrochlorofluorocarbons (HCFCs; purple circles). The latter are likely replacements for the former, at least in the short term. The results are derived from 1D computer model studies. The area of each circle is proportional to the species's atmospheric lifetime. The calculated ODPs and GWPs presented here are both referenced to those of CFC-11, which are each assigned a value of 1.0. Note that the replacement products under discussion all have much lower ODPs than the CFCs, but some have moderately high GWPs.

of ozone that is expected to be destroyed by the emission of a quantity of gas over the gas's entire atmospheric lifetime, defined relative to the ozone destruction due to CFC-11;

$$ODP_x = \frac{\text{Time-integrated global O}_3 \text{ loss due to } x}{\text{Time-integrated global O}_3 \text{ loss due to CFC-11}} \quad (6.1)$$

The lower the ODP_x value, the better suited is x as a CFC replacement. An advantage of this approach is that, since the ODP is a relative measure, it is likely to have greater reliability than would be the case if absolute ozone depletions were being compared.

Anthropogenic Enhancement of the Greenhouse Effect

One-dimensional models are also commonly used to investigate the greenhouse effect, the overall atmospheric radiation budget, and the degree to which emissions from human activities could perturb these critical processes. It

was pointed out in Chapter 2 that the basic source of energy for most terrestrial life, and certainly for climate, is the absorption of solar radiation. If the climate system is to be in equilibrium, the incident solar radiation that reaches Earth must be balanced by the thermal radiation that Earth sends out into space. Much of the current concern over the impact of human activities on climate arises from the fact, as we have also seen, that those activities have been altering the amount of absorbed and emitted radiation that is retained in the Earth–atmosphere system, as well as changing the hydrologic cycle of the planet. To assess the magnitude of these impacts, scientists need to study the interactions between anthropogenic and natural variations in climate and its driving forces.

Many different trace gases have the potential to alter Earth's climate. The most important is water vapor, whose atmospheric concentrations cannot be directly influenced by human activity to any significant degree. Next in importance is carbon dioxide, whose effect is calculated to be large over the next century. Almost equally important during the past two decades has been a group of other gases, especially methane, nitrous oxide, and the CFCs, with long atmospheric lifetimes and absorption characteristics that allow them to interact with terrestrial infrared radiation. It is important to realize that the combined effect of the anthropogenic increases in these gases is enhanced by about a factor of 2 because of the increase in water vapor that can be expected to accompany any temperature rise at the Earth's surface as a consequence of increased evaporation from the oceans.

An increase in the concentration of a greenhouse gas initially decreases the flux of long-wave terrestrial radiation to space as the gas traps more of that radiation in the troposphere (see the diagram on page 14 of Chapter 2). The effect will be a temperature rise at Earth's surface, the amount depending in part on related processes such as changes in water evaporation rates and cloudiness.

When comparing the heating powers of different greenhouse gases, scientists try to estimate, for each gas, the overall reduction in infrared radiation leaving Earth per unit increase in the gas's atmospheric abundance. The results are often expressed as global warming potentials (GWPs), founded on a line of reasoning and computed with tools similar to those for ODPs. GWPs assess the relative abilities of a given quantity of x (compared to an equal quantity of carbon dioxide) to diminish the terrestrial infrared radiation that escapes to space over an extended time period (often a decade or a century is used for calculation), thus leading to a heating of the Earth system. Although the actual calculation of ODPs and GWPs is rather intricate and requires the use of computer models, the results have proven to be effective tools in the political arena as international agreements are developed for ameliorating global atmospheric change.

GWP analyses reveal that the greenhouse effect of methane is about 25 times larger than that of carbon dioxide on a molecule-to-molecule basis. The reason is that there is already so much carbon dioxide in the atmosphere that in many spectral regions the radiation absorption is almost complete, and thus most of the effect of added carbon dioxide occurs only at the edges of its absorption bands. In contrast, several less abundant greenhouse gases, including methane, absorb wavelengths that currently are being captured less effectively, and each added molecule provides significant new absorption capabilities. For example, increasing the atmospheric carbon dioxide concentration (now at 360 ppmv) by 250 ppmv would increase greenhouse warming by about 3.3 watts per square meter ($W\ m^{-2}$), whereas increasing methane (now at 1.6 ppmv) by 3 ppmv would raise it by 1.3 $W\ m^{-2}$. The results would be even more marked for increases in CFC-11, CFC-12, and nitrous oxide.

Computer models able to incorporate the radiative properties of greenhouses gases along with specific emission scenarios can also be used to calculate likely temperatures for Earth's surface at different times in the near and intermediate future. In the illustration on page 120 several such calculations based on projections of future atmospheric emissions have been grouped to show ranges of possible low, intermediate, and high temperature increases throughout the next half century. The spreads in the ranges reflect current uncertainties about the degree to which climate is sensitive to changes in greenhouse gas concentrations, the degree and rate with which the oceans can absorb increased heat from the air, and other factors. It seems very likely, however, that should no significant restraint be placed on the emissions of greenhouse gases, the average global temperature over the next half century

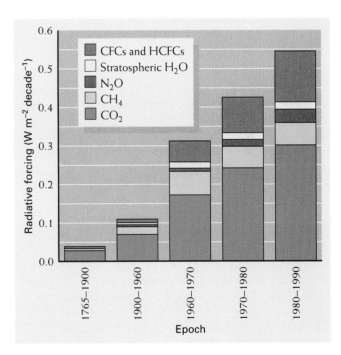

The degree of radiative forcing produced by selected greenhouse gases in five different epochs. Until about 1960, nearly all the forcing was due to CO_2; between that time and today, the other greenhouse gases have combined to nearly equal the CO_2 forcing.

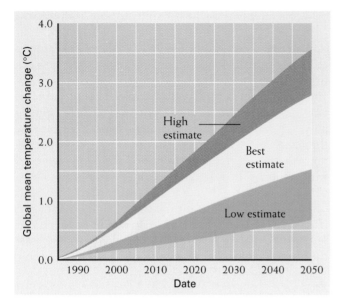

The annual mean surface air warming for computations that assume "business as usual" rates of greenhouse gas emissions but consider a variety of climate sensitivities, oceanic heat uptake rates, and other variables.

will indeed increase to somewhere within the ranges indicated. Other calculations have indicated that even if emissions are severely limited, the average global temperature will rise by at least 0.7 to 1.5°C by 2050.

The source of greatest uncertainty in even the most sophisticated climate change predictions is thought to be the treatment of clouds in the models. Higher temperatures bring more water vapor into the atmosphere, which in turn results in the formation of more clouds, and clouds affect climate in two ways. On the one hand, they reflect back to space part of the Sun's radiation before it reaches the ground, and thus they have a cooling effect. On the other hand, they trap upward-moving heat radiation and thus they also have a warming effect. The net result of these two processes depends on the clouds' altitudes. High (cirrus) clouds reduce outgoing heat radiation more than they reflect incoming solar radiation, thus adding energy to the Earth–atmosphere system; low clouds have the

opposite effect. The models must therefore predict the rates at which changes in temperature will produce clouds at different altitudes and seasons, all the while dealing with the complexities of cloud-droplet nucleation around particles, altitude and humidity changes over different continents and subcontinents, and the different possible air-parcel motions under different planetary temperature conditions.

The possible global average temperature increase of one or two degrees centigrade over the next few decades may not sound very large until we compare it with past climate oscillations and consider what their effects have been. For example, the average temperature during the Little Ice Age of 1400 to 1650 or thereabouts was only about 1.0°C cooler than the present. Just that small amount of cooling necessitated significant changes in agricultural practice and habitation—proof that global temperature changes of that slight a magnitude are definitely of major importance. In addition, should a warming of 1 or 2°C occur, the planet would reach an average temperature that has probably not been seen in 120,000 years. Computer models estimate that among the results of such warming would be an increase in global mean sea level of perhaps 20 cm by the year 2030 and perhaps 45 cm by 2070; this will occur because the ocean water will expand slightly as it is heated and because a warmer climate will cause increased melting of mountain glaciers. Such sea level increases would produce major disruptions for the large fraction of the world's population living in coastal regions, especially the less developed nations in Southeast Asia; for example, in Bangladesh, most of the population lives at or near the water's edge. Finally, suppose that the higher-temperature warming scenarios occur, a prospect as likely as the lower ones. A change of 3°C or more would be comparable to the temperature change that occurred between the last major ice age and the present, a transition that took place over several thousand years. In contrast, humankind would have produced, within a period of only a century, the warmest climate to have existed in millions of years. We have only a vague idea of how our natural systems would respond. Should we thus continue the "business as usual" path and accept the risks?

In addition to predicting how changes in the concentrations of today's greenhouse gases might affect future cli-

Chapter Six

mates, it is worth asking whether other gases might be involved in future climate change. This inquiry leads us to consider the basic defining features of a greenhouse gas. What characteristics do the most important greenhouse gases (H_2O, CO_2, CH_4, N_2O, $CFCl_3$, and CF_2Cl_2) have in common? First, they are unreactive toward the hydroxyl radical (HO·), the principal atmospheric removal initiator. (Methane seems at first to be an exception, but because its C-H bonds are the strongest of all possible C-H bonds, as a consequence of its structural symmetry, the CH_4 + HO· reaction rate is very slow.) This requirement, for low or negligible atmospheric reactivity, eliminates all other organic molecules from consideration as potential greenhouse gases, because their exposed hydrogen atoms are readily attacked by HO·. The same argument holds for hydrogenated inorganic molecules, such as hydrogen sulfide. Also eliminated are inorganic acid precursor molecules, such as nitric oxide, nitrogen dioxide, and sulfur dioxide. These molecules form addition compounds with HO· that are easily removed from the atmosphere by rainfall. A second characteristic of greenhouse gases is that they are not dissociated by solar photons in the visible and near-ultraviolet portions of the spectrum. This requirement eliminates as potential greenhouse gases any molecules with O-O bonds, such as hydrogen peroxide (H_2O_2). (Ozone seems at first to be an exception, but the oxygen atom liberated by the photolysis reacts with molecular oxygen and reforms ozone with such a high probability that photolysis is not an important sink.) Third, greenhouse gas molecules are insoluble, or nearly so, in water. This requirement eliminates as a potential greenhouse gas any organic molecule with an −OH group, such as formic acid (HCOOH), and many inorganic species, such as hydrochloric and nitric acids and ammonia. These three requirements are basically the characteristics needed if a molecule is to have a long atmospheric lifetime. The fourth and final requirement is that greenhouse gas molecules must have chemical bonds that absorb radiation within the 8 to 13 μm window left by carbon dioxide and water. The gases O_3, CH_4, N_2O, the CFCs, CF_4, and C_2F_6 meet this requirement. We are fortunate that these four criteria are quite restrictive; it seems unlikely that any significant additions will ever be made to the current list of greenhouse gases.

Atmospheric Effects of Supersonic Aircraft

The 1D model readily probes differences in cloud formation between the troposphere and stratosphere, or between ozone chemistry at the ground and aloft, but it is unable to study another factor everyone knows to be important: the changes of chemistry and climate with latitude. Solar radiation obviously is more intense at the equator than at the poles, emissions are greater in the Northern Hemisphere than in the Southern, and so forth. To capture these complexities, the 2D model, shown in the figure on page 124, retains the vertical layers of the 1D model and adds a dimension in latitude. This model can be thought of as laying a grid on the sky and computing results separately for each space within the grid.

A closeup view of one of the grid spaces reveals the consequences of this added dimension. Each grid space behaves as a box model in that it receives inputs of emissions, radiation, and air flows, and undergoes chemical reactions, cloud formation, and so forth; but it also serves as a source of inputs for grid spaces adjacent in both latitude and altitude. For the calculation to be valid, no mass or energy can be gained or lost when all the grid space results are summed up: with all the transfers, there must still be just as many oxygen atoms, as many nitrogen atoms, as many chlorine atoms as when the calculation began, and all the energies must be accounted for.

Computer size and speed rapidly become factors in 2D model design. Whereas the 1D model dealt with 10 or 15 layers simultaneously, the 2D model deals with perhaps 90 to 270 grid spaces (10 to 15 layers in altitude by 9 to 18 sectors in latitude). Within each of these grid spaces, numerous chemical reactions and physical processes must be simulated. As a consequence, efficient computational techniques become as important to success as an accurate representation of chemistry and physics.

While the 2D model avoids the 1D model's assumption that every point within an altitude layer is identical to every other, it must nonetheless retain the simplification that at a given altitude and latitude, all points are chemically and meteorologically the same regardless of longitude. This is equivalent to an implicit assumption

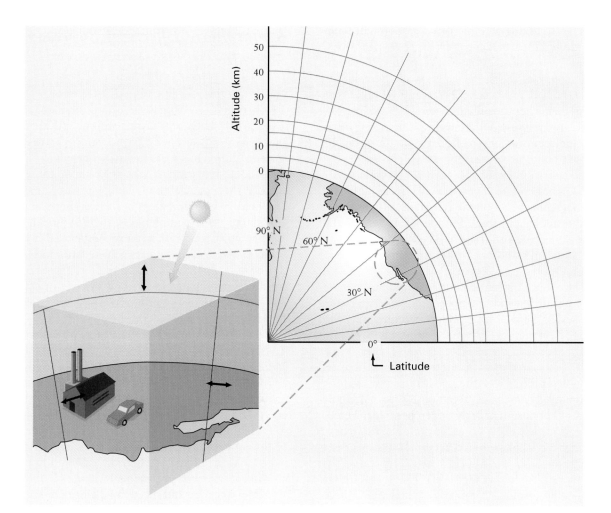

A schematic diagram showing the features of a 2D model of atmospheric chemistry. Each grid space is identified by a particular range in latitude and altitude; it receives solar radiation from above and interchanges air containing gases, particles, and water vapor with its neighboring grid spaces.

that variations in chemical species concentrations in the longitudinal (west to east) directions are much smaller than those in the vertical and north to south directions. The neglect of longitudinal variations will invariably introduce deviations from reality, especially at lower altitudes, where the influence of chemical and biological processes as sources and sinks of trace gases at the Earth's surface is large. Two-dimensional models have, therefore, mainly been used for stratospheric studies.

An example of a stratospheric 2D model calculation is the work of Malcolm Ko and his colleagues at Atmospheric and Environmental Research, Inc., in Cambridge, Massachusetts, in which they addressed the question, To what extent would the proposed large fleet of supersonic aircraft perturb the stratospheric ozone layer in the year 2015? As its input conditions, the computation required estimates of exhaust emissions from present subsonic and proposed supersonic aircraft engines and predictions of

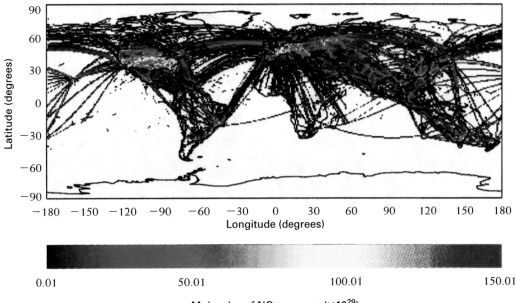

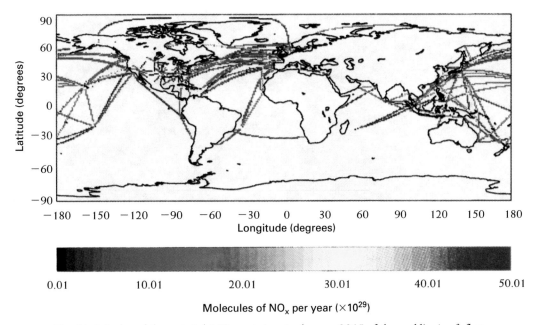

Graphical display of the projected NO_x emissions in the year 2015 of the world's aircraft fleet. (Top) Emissions below 13 km altitude, primarily due to subsonic aircraft. (Bottom) Emissions above 13 km altitude, primarily due to supersonic aircraft. The spatial resolution of the inventory is 1° latitude by 1° longitude. Similar plots could readily by generated for other altitudes or other emittants.

Predicting the Near Future

the number and route of supersonic flights, and, as a base condition, a projection of the atmosphere's composition in 2015 assuming the absence of the supersonic fleet. An engine exhaust species of particular concern was nitrogen dioxide (NO_2), a known participant in ozone catalytic reaction cycles.

To construct the emissions inventory for the world's subsonic and supersonic aircraft now and in 2015, a group of researchers and aviation experts assembled information on scheduled airline flights, typical flight patterns, and numbers of military and private aircraft. The emissions from each of the engine types at different temperatures and pressures were then measured or estimated. Next, given average sequences (taxi, climb, cruise, descend, and land), emissions were computed for a typical flight. These computations were then extended to represent global sub-

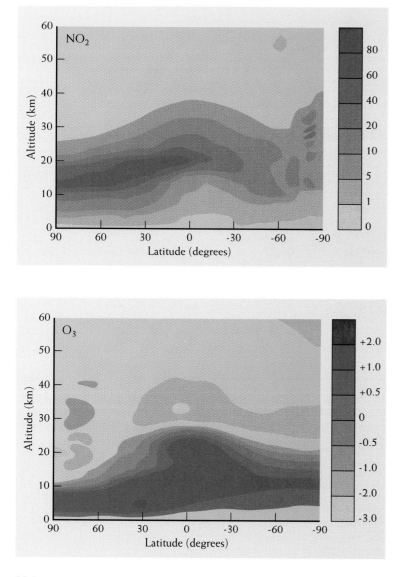

A 2D model's assessment of the impact of a fleet of supersonic aircraft on stratospheric ozone. The top panel shows the percent enhancement of NO_2 (a major aircraft emittant) relative to conditions expected in the absence of such a fleet. The bottom panel shows the percent ozone change as a function of altitude and latitude.

sonic and supersonic flight types and frequencies and developed into an inventory for several emitted species as a function of altitude, latitude, and longitude.

Using this emissions inventory as a foundation, Ko's global 2D model then added the standard characteristics of the atmosphere, its chemical reactions, and its nonaircraft sources and sinks. The results of the computation (shown in the graphs above and preliminary as of this writing) turned out to be counterintuitive, a consequence of the presence of aerosol particles. In a process reminiscent of the central role played by particle surfaces in creating the Antarctic ozone hole, most of the emitted NO_2 was converted to dinitrogen pentoxide (N_2O_5) by

$$NO_2 + O_3 \rightarrow NO_3 + O_2 \qquad (6.2)$$

$$NO_2 + NO_3 + M \rightarrow N_2O_5 + M \qquad (6.3)$$

and the N_2O_5 was transformed to nitric acid on the surface of water-rich stratospheric aerosol particles

$$N_2O_5 + H_2O_{(particle)} \rightarrow 2\ HNO_3 \qquad (6.4)$$

Because of the enhanced conversion of NO_2 to HNO_3, NO_2 is no longer as abundantly available for reaction with atomic oxygen and subsequent ozone destruction by the reactions

$$NO_2 + O \rightarrow NO + O_2 \qquad (6.5)$$
$$\text{and}$$
$$NO + O_3 \rightarrow NO_2 + O_2 \qquad (6.6)$$

As a consequence, even though stratospheric NO_2 is enhanced by as much as 40% in and near the principal corridors anticipated for supersonic flight, the decrease in column ozone there is currently predicted to be less than 1.5%.

The Stability of Atmospheric Oxidizers

In Chapter 3 we discussed the vital importance both of ozone and (especially) of the hydroxyl radical in oxidizing a very large number of environmentally detrimental emittants and thus removing them from the atmosphere. Among the questions of interest for the near future, therefore, is whether global increases in emissions could possibly deplete the atmosphere's supply of those crucial oxidizers. If this were to happen, the air quality in urban boundary layers as well as in the more distant free troposphere would deteriorate markedly.

As far as scientists can tell, there is no reason to expect that ozone concentrations will decrease with increased emissions. In fact, we presented evidence in Chapter 5 that tropospheric ozone levels have instead been *increasing* for several decades. This increase is widely believed to be a consequence of the production of NO during the high-temperature combustion of fossil fuels and its oxidation to NO_2 by hydrocarbon intermediates, followed by

$$NO_2 + h\nu \rightarrow NO + O \qquad (6.7)$$
$$O + O_2 \rightarrow O_3 \qquad (6.8)$$

As for $HO\cdot$, because its concentrations are so low, they can be measured only with great difficulty. As a consequence, data are few and $HO\cdot$ concentration trends have not been established from measurements of the species made over long periods of time. However, an indirect method has been devised that makes use of data from a global monitoring network formed to study the concentrations of the solvent methyl chloroform (CH_3CCl_3). Other than transport to the stratosphere, the only mechanism that removes methyl chloroform from the troposphere is reaction with $HO\cdot$. Thus, if the locations and magnitudes of the emissions are well known and if enough methyl chloroform concentration measurements are available, computer studies can estimate the concentrations of $HO\cdot$ needed to explain those observed abundances. The results of these calculations do not yet have a high degree of reliability, but they do suggest that the global $HO\cdot$ concentration has been essentially constant for the past 15 years.

Variations in Surface Ozone

The crowning achievement of computational atmospheric science is the three-dimensional (3D) model, in which variation in longitude is added to the 2D concept. The individual entity in the computation, the grid space, now becomes a volume representing a specific geographical area and a specific altitude range. The additional requirements on the 3D model builder are daunting. She or he must accurately represent the major air flows throughout the entire atmosphere—the Hadley circulation, the jet streams, the transfers between stratosphere and troposphere—because interchanges of air must occur among all the grid volumes: up and down, east and west, north and south. Emissions of pollutants must be specified with reasonable precision for every country of appropriate size, on every continent. Cloud formation over oceans and over the land must be simulated, and radiant energy transfer must be accurately reproduced. In the creation of a 3D model, compromise quickly becomes the order of the day because the number of grid volumes to be simultaneously treated may be as many as several thousand.

For many problems of interest in atmospheric chemistry, a full 3D treatment is beyond today's computer resources, yet the problems are too complex to be studied with a 2D model. An intermediate modeling approach that has proved useful in some of these cases is the *decoupled* 3D model. In this technique, researchers use a meteorological 3D model to calculate winds, temperature, water vapor, and cloud distributions for a chosen geographical region (in some cases, the entire globe) and a time of year. Those results are then stored to be used as input information for a 3D atmospheric chemistry model, a tactic that simplifies the computations and allows more time and computer storage space for doing each problem. The decoupled approach works for situations where the changing chemical concentrations will not significantly alter the climate itself. This is the case for many tropospheric applications in which calculations are performed for limited time periods of perhaps a few years or less; such a model, however, would not be valid if enough greenhouse gases were added to alter the atmospheric wind and temperature structure. For the stratosphere, decoupled models are, at best, only approximate descriptions of reality because the most important feature of the stratosphere—the ozone concentration—determines the heating and thus the circulation of the air there.

The advantage of a decoupled 3D model is its ability to simulate photochemical processes in much more detail than a coupled one. Because decoupled models can perform very efficiently, they can be used to study various alternative scenarios. For example, a decoupled model simulated surface ozone concentrations for the preindustrial era and for the present, concluding that surface ozone concentrations may have more than doubled over the past century in the northern hemisphere, mostly as a result of growing industrial emissions of NO_x, the catalyst for the production of tropospheric ozone. This result is in approximate agreement with the observational data shown in the illustration on page 102.

Models of Urban Smog

For almost a quarter of a century, model builders have been simulating, with growing sophistication, the chemistry and meteorology involved in smog formation and dispersal. Among the more recent efforts is a model developed at Carnegie-Mellon University by Greg McRae (now at the Massachusetts Institute of Technology) and Armistead Russell to study ozone formation in the Los Angeles basin. Their model is a full 3D approach, using a horizontal grid volume spacing of 5 km in the north-to-south and east-to-west directions and a variable vertical resolution, creating a mesh of 12,000 grid volumes. Among the inputs are spatially distinct inventories of emissions from automobiles, factories, and other sources, resolved by time of day, as well as typical meteorological conditions for the basin and its inflowing and outflowing air currents.

McRae and Russell's work has confirmed earlier results from lower-dimensional models indicating that both hydrocarbon and nitrogen oxide emissions are crucial to ozone control and that both chemistry and transport are vital factors in determining the geographical extent of the smog problem in Los Angeles. In addition, it identifies the problem areas in the city much more precisely than ever before and uses advanced computer graphics to demonstrate the spatial distribution of the emittants and their chemical products. These results, and those of other investigators, have given convincing proof of the abilities of computer modeling tools to provide reliable guidance toward improving the air quality of many of the planet's large urban complexes.

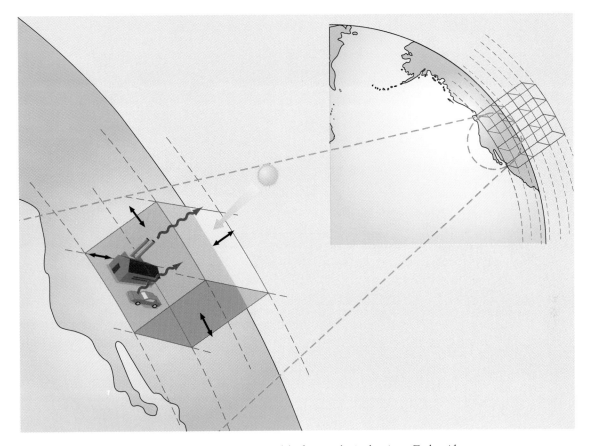

A schematic diagram showing the features of a 3D model of atmospheric chemistry. Each grid volume is identified by a particular range in latitude, longitude, and altitude; it receives solar radiation from above and interchanges air containing gases, particles, and water vapor with its neighboring grid spaces.

Detailed Modeling of the Climate System

The most comprehensive atmospheric models, the full 3D global models, are able to calculate the varying concentrations of one or more chemical species in three spatial dimensions and one time dimension simultaneously. The necessary equations are extremely complex and enormously difficult to solve, and the satisfactory construction of models with the requisite capabilities is only now coming within the grasp of the atmospheric science community (and then only with use of today's largest and fastest computers). An example of these research tools and their application is the use of general circulation models (GCMs) to simulate global climate.

A GCM takes the approach of McRae and Russell, but extends it over the entire planet and from the surface to well into the stratosphere. Its goal is not short-term air quality prediction, but prediction of global climate decades

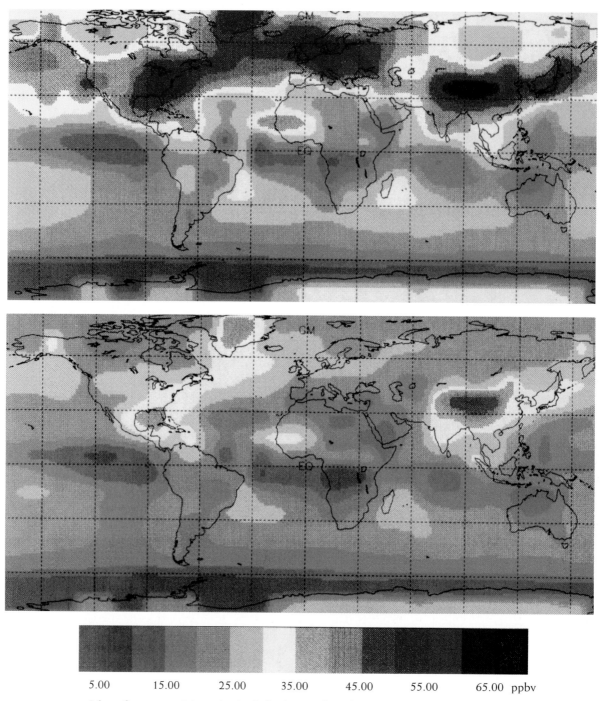

July surface ozone mixing ratios (top) for the preindustrial era and (bottom) for the present, as computed by the decoupled 3D global transport and chemistry model of Paul J. Crutzen and colleagues.

A visualization of smog and its precursors over the Los Angeles basin, generated by the 3D computer model of Gregory McRae of the Massachusetts Institute of Technology and Alistair Russell of Carnegie-Mellon University.

or centuries hence. In one of the most relevant of these applications, Syukuro Manabe and his colleagues at the Geophysical Fluid Dynamics Laboratory in Princeton, New Jersey, studied the temperature effects that may occur in specific geographical regions and around the globe should greenhouse gas concentrations continue to increase. These researchers interposed different concentrations of carbon dioxide into their 3D global climate model and then calculated changes in temperature across their model's entire grid. The illustration on page 132 shows a sample of their work: the prediction of temperature changes that would follow a quadrupling of the atmospheric carbon dioxide concentration, a condition that some think could develop during the latter part of the twenty-first century.

The model predicts substantial geographical variability in the annual average temperature increases. Near the equator, the anticipated warming trends are 2 or 3°C. At the poles, however, the changes would be much larger: up to almost 9°C at the North Pole. Because a lot of Earth's water is stored at the poles in ice caps, such changes could result in much increased ice cap melting and a significant rise in sea level.

Other 3D models, designed to investigate such parameters as soil moisture, precipitation, continental runoff rates, and other climate-related factors, have also predicted large geographical climate variations. An example is the GCM of David Rind, James Hansen, and their colleagues at the Goddard Institute for Space Studies, a NASA facility in New York City. These workers, using the climate-change projections from a trace gas scenario in which radiative greenhouse forcing continues to grow at an exponential rate, pinpointed the regions on Earth that would experience drought, the calculation treating four model years. For 1969, their model (and supporting data that were examined) shows an equal occurrence of wet and dry regions, with extreme occurrences randomly distributed and infrequent. By 1999, very dry weather conditions occur over some tropical and subtropical land masses. In 2029, the dry regions have expanded, pushing into a number of midlatitude areas and increasing in intensity. Extreme drought covers most midlatitude locations, and extreme flood conditions are found at the highest latitudes by 2059. Rind and his coworkers point out that if the trace gas emission increases occur at a lower rate than they have assumed, or if the climate sensitivity

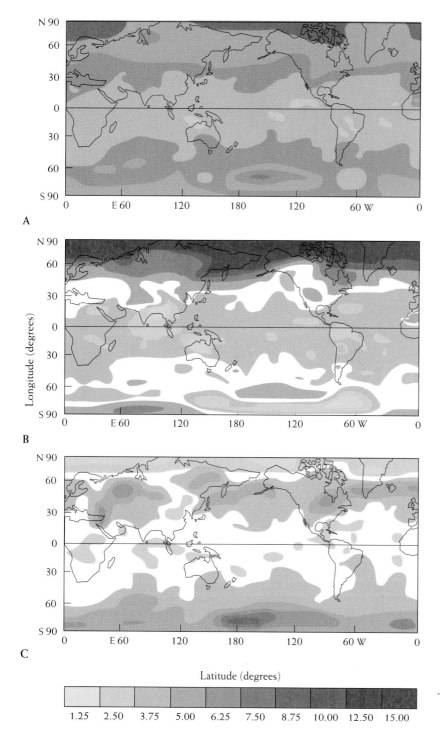

A

B

C

Latitude (degrees)

1.25 2.50 3.75 5.00 6.25 7.50 8.75 10.00 12.50 15.00

The geographical distribution of mean surface air warming (in °C) for a computation that assumes an atmospheric CO₂ concentration four times the present value. (a) Annual; (b) December through February; (c) June through August.

Chapter Six

The drought-parched bed of the Aral Sea in south-central Asia, normally the fourth-largest inland body of water on the planet.

is less, the effects will be modified accordingly. Nonetheless, if climate changes even approaching their results become a reality rather than just a prediction, the impact on plants, animals, and humans would be very significant indeed, so the model's prediction should be taken as a serious warning.

Any regional or global temperature change is likely to mean the redistribution of various types of biota. One dendrological study considers the possible fate of the eastern hemlock tree if temperature conditions predicted by GCMs should come to pass. Because the rates of temperature change anticipated for the coming centuries are so

A dust storm clouds the sky in Elkhart, Kansas, May 21, 1937. The United States Great Plains, termed the Dust Bowl because of extended drought in the 1930s, are now marginally stable as a result of improved farming practices and a return to moderate rates of precipitation.

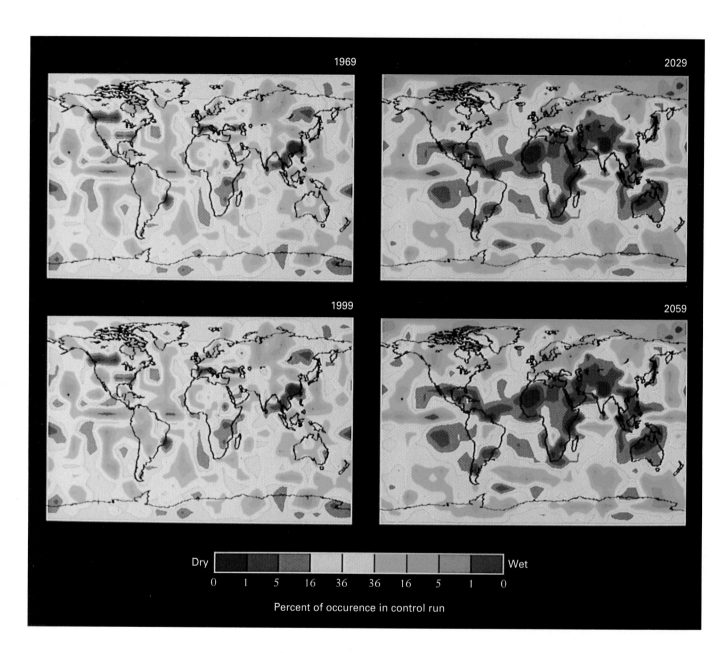

1969

2029

1999

2059

Dry ▮▮▮▮▯▯▮▮▮▮ Wet

0   1   5   16   36   36   16   5   1   0

Percent of occurence in control run

The occurrence of drought for the period June through August of four specified years in a global climate model simulation. The color scale is keyed to the frequency of occurrence of drought in a 100-year control run that had a 1958 atmospheric composition.

Regional Climates of the Year 2030

*T*he results of several recent GCM research efforts were combined and summarized in 1990 by the Intergovernmental Panel on Climate Change (IPCC). This organization, a joint project of the World Meteorological Organization and the United Nations Environment Programme, was established to assess the scientific information related to climate change. Several hundred working scientists from more than two dozen countries have been involved in this continuing project, producing updated reports on a periodic basis. Among the topics studied by this group were possible future climate changes in five geographical regions. For each region, the IPCC estimated changes in temperature, precipitation, and soil moisture between 1990 and 2030, scaled to be consistent with the IPCC "best estimate" of a global mean warming of 1.8°C by 2030. The scientific community's confidence in the regional results is not particularly high, because the global models are not able to take all potentially relevant information into account and do not have enough spatial resolution to make accurate predictions about precipitation and soil moisture. However, the calculations are more than sufficient to be a strong indicator that substantial regional changes may occur; the possibility of climate change is thus an issue of grave concern on a regional as well as a global basis. The predictions for the five regions are as follows:

Central North America (35° to 50° N, 85° to 105° W) The warming varies from 2 to 4°C in winter and 2 to 3°C in the summer. Precipitation increases up to 15% in winter but decreases by 5 to 10% in summer. Soil moisture increases by 5 to 10%.

Southern Asia (5° to 30° N, 70° to 105° E) The warming varies from 1 to 2°C throughout the year. Precipitation changes little in winter and generally increases throughout the region by 5 to 15% in summer. Summer soil moisture increases by 5 to 10%.

Sahel Desert, Africa (10° to 20° N, 20° W to 40° E) The warming ranges from 1 to 3°C. Area mean precipitation increases, and area mean soil moisture decreases marginally in summer. However, throughout the region there are areas of both increase and decrease in both parameters.

Southern Europe (35° to 50° N, 10° W to 45° E) The warming is about 2°C in winter and varies from 2 to 3°C in summer. There is some indication of increased precipitation in winter, but summer precipitation decreases by 5 to 15%, and summer soil moisture by 15 to 25%.

Australia (12° to 45° S, 110° to 155° E) The warming ranges from 1 to 2°C in summer and is about 2°C in winter. Summer precipitation increases by around 10%, but the models do not produce consistent estimates of the changes in soil moisture. The area averages mask large variations at the subcontinental level.

rapid compared to the tree's generation time, the hemlock forests may not be able to relocate successfully; as a consequence, they are likely to disappear from much of their current range, while large portions of their potential range will remain unoccupied for a long period of time. In other words, although the overall effects of future climate changes may prove to be beneficial for some ecosystems and some nations several centuries from now, the expected rapidity of those changes will present major problems during the intervening time, because environmental adaptation will be too slow to transfer healthy ecosystems intact from one geographical location to another.

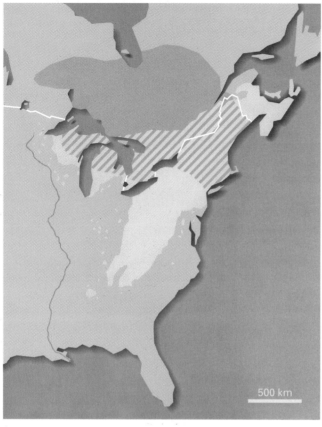
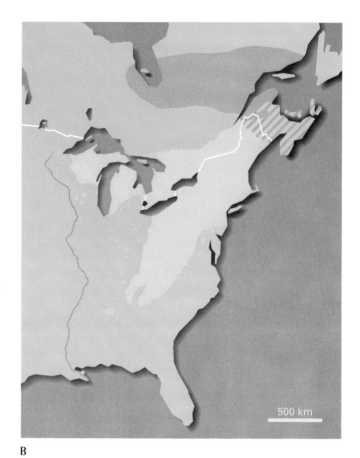

A

B

The present and future geographical range of the eastern hemlock in northeastern North America. The yellow area indicates the present range, the green area indicates the extent of the range in about the year 2080, and the striped area is where the two ranges overlap. The diagram on the left reflects temperatures calculated by the computer model at the Goddard Institute for Space Studies, New York, NY. The diagram on the right relies on temperatures generated by the computer model at the Geophysical Fluid Dynamics Laboratory, Princeton, New Jersey. In both cases, the hemlock range would initially be expected to be reduced to the striped area.

The Cooling Effects of SO₂ Emissions

We pointed out earlier that the anthropogenic emissions of sulfur—usually in the form of SO_2—now quantitatively surpass the sulfur emissions from nature. Once emitted, the SO_2 is efficiently transformed to sulfuric acid (H_2SO_4) by atmospheric reactions:

$$SO_2 + HO\cdot \rightarrow HOSO_2\cdot \qquad (6.9)$$

$$HOSO_2\cdot + O_2 \rightarrow SO_3 + HO_2\cdot \qquad (6.10)$$

$$SO_3 + H_2O \rightarrow H_2SO_4 \qquad (6.11)$$

Because sulfuric acid molecules efficiently attract water vapor and condense to form sulfate particles capable of scattering light, their presence in the air can greatly reduce visibility in polluted regions, as well as cause delete-rious effects in the biosphere when they are incorporated into droplets and fall as acidic precipitation. For those reasons, many developed nations are reducing sulfur emissions by using fuels with low levels of sulfur and by extensive control of emissions from industrial and combustion processes.

The impact of sulfur on a global rather than on a regional scale, however, has not been entirely detrimental, at least from a short-term perspective. The reason is that sulfate particles in the air have a cooling effect on the atmosphere because they enhance the backscattering to space of solar radiation, thus reducing the amount of solar energy reaching and being absorbed by Earth's surface. Recently, the potential significance of this effect has been investigated in a collaborative theoretical effort by scientists from the United States, Sweden, and Germany, who used a 3D chemistry-climate model to calculate the spatial distribution of sulfate particles from both natural and anthropogenic sulfur emissions and then computed the influence of those particles on the atmospheric radiation budget.

The maps on the following page compare the radiative effects of the anthropogenic sulfur emissions alone to the effects of the anthropogenic plus natural sulfur emissions combined. The impact of the anthropogenic emissions is clear, especially over and downwind of those parts of the world where the bulk of the sulfur dioxide emissions occur. The calculations also show, however, that the influence of the anthropogenic emissions is restricted almost entirely to the northern hemisphere; which is not surprising in view of the sulfate particle's short atmospheric lifetime before being removed by precipitation.

The magnitude of the backscattering at different geographical locations is also of interest. Values largely due to anthropogenic emissions can be as much as 4 W m^{-2} over southeastern Europe, 2 W m^{-2} over the southern United States, and about 1 W m^{-2} for the entire Northern Hemisphere. An increase of 1 W m^{-2} is just about the magnitude of radiative forcing caused by the increase of carbon dioxide in the atmosphere since preindustrial times. However, the influence of sulfur is opposite that of carbon dioxide. On a regional basis, at least, the industrial and combustion-related sulfur dioxide emissions may thus have counteracted an appreciable portion of the warming from anthropogenic greenhouse gases. Notwithstanding

this interesting idea, now being tested by analyses of temperature records and by field experiments of various types, the authors of the original collaborative study point out that the sulfur dioxide and carbon dioxide interactions with radiation are not uniform over all parts of the globe and that the anticipated effects differ in each hemisphere and at different locations within each hemisphere.

A final point to consider regarding sulfur is that the effects of sulfate particles on climate may also involve increases in cloud albedo, which likewise favors climate cooling, because sulfate particles are extremely efficient cloud-condensation nuclei. Thus, an increase in sulfur dioxide emissions followed by the formation of sulfate aerosol could lead to higher concentrations of cloud droplets at the same atmospheric liquid water concentration and thus greater droplet surface areas. The result is whiter clouds that are more efficient scatterers of solar radiation.

Atmosphere–Ocean Circulation Models

The extensive interactions of the atmosphere and the oceans can be illustrated by a few obvious examples: ocean mixing and chemistry determine the transport of atmospheric carbon to the deep ocean and its sequestration in sediments, thus strongly influencing the global carbon cycle; ocean currents are responsible for much of the transport of heat from low to high latitudes, thus playing a major role in climate; and sulfurated gases emitted from the sea surface play major roles in the rate of cloud formation. Hence, global atmospheric chemistry and climate cannot possibly be assessed without a detailed understanding of ocean currents, ocean chemistry, and the response of the oceans to external stimuli such as temperature change.

Although oceanic model building is a specialty in itself, it has certain commonalities with its atmospheric analogue. Oceans and atmospheres are both fluid systems influenced by the changing concentrations of trace species, and ocean models, like atmospheric models, begin with the equations of motion and of mass balance for chemical and physical properties (such as momentum, heat, and salinity). To some extent, an ocean model can be pictured as an inverted atmospheric model, with high resolution (that is, many thin layers) near the surface and low resolution (that is, a few thick layers) at great depths.

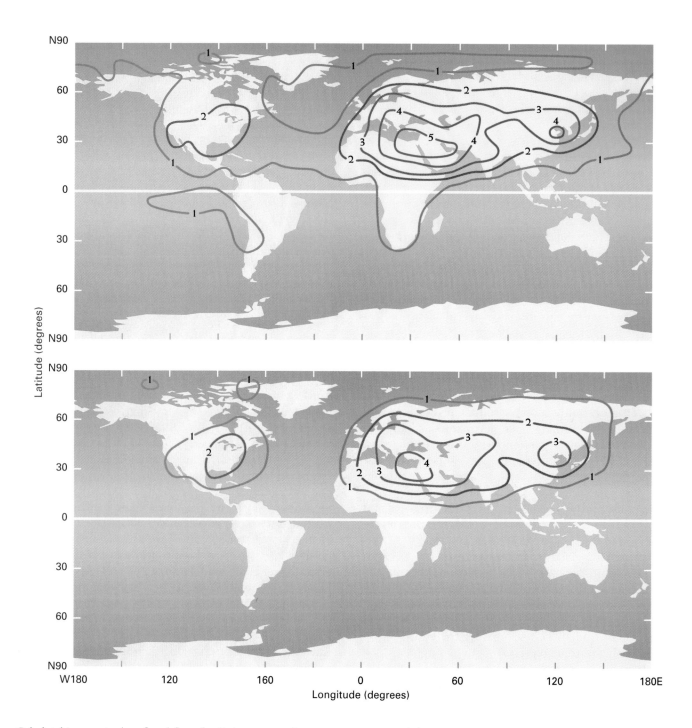

Calculated increase in the reflected flux of radiation to space (in watts per square meter) due to airborne sulfate particles. In the top diagram, both anthropogenic and natural sources of sulfur are included in the calcualtions. In the bottom, anthropogenic sources only are included.

Chapter Six

Just as the atmospheric models must take account of the way air moves around mountain ranges and other obstacles, ocean models must deal with the topography of the ocean floor and the continents. The boundary conditions include input from rivers, loss to evaporation, response to surface winds, and a host of other factors. In many cases, a sink for an atmospheric model is a source for an ocean model, and vice versa.

Perhaps the best example of the abilities of current ocean models is the performance of a 3D model constructed by Albert Semtner of the Naval Postgraduate School and Robert Chervin of the National Center for Atmospheric Research, both in the United States. Their model has 20 vertical levels and a horizontal grid size of 1/6° by 1/6°, and runs on modern parallel supercomputers. Just as atmospheric model calculations predict jet streams and other flow features, so the ocean model results predict large and small flow features that can then be compared with actual observations. The results of this computer model for a layer 160 m below the surface of the world's oceans match the principal underwater currents—in particular, the "conveyer-belt circulation" that we discussed in Chapter 2. Both the flow inferred from shipboard measurements and the flow calculated by the computer model have a warm, shallow component that moves from east to west across the central Pacific Ocean. Many of the details of this flow have been mapped by sensors towed underwater from ships. The model describes more localized flow structures equally well, such as those around continental boundaries and undersea mountain barriers. The detailed results serve as an oceanic reference point for future studies of the interchange of heat, water, and gases between ocean and atmosphere.

Computer models of the oceans and of the atmosphere have traditionally been formulated separately, but in the real world the behavior of the atmosphere and oceans is so strongly coupled that their interactions are important features to capture. The time scales for mixing and transport are very different, however, that for the atmosphere being a few days to a year, that for the oceans decades to centuries. Three-dimensional approaches that attempt to encompass these factors in a single calculation, such as the model of Syukuro Manabe and his colleagues at the Geophysical Fluid Dynamics Laboratory, Princeton, New Jersey, are thus perhaps the most ambitious of today's

Earth system models. In order to fit the model into the computer, the investigators simplified the oceanic features significantly. Manabe and his coworkers found that when changes in Earth's radiation budget are simulated, sharp responses are seen in the model's temperature and sea level results. For example, the model predicted that a quadrupling of the present atmospheric carbon dioxide level would lead to a temperature increase of 3 to 4°C per century and a rise in sea level of 30 to 40 cm per century. As we mentioned earlier, much of Earth's population would find it inordinately difficult to adjust to such changes.

Potentially even more significant than temperature and sea level changes in response to a quadrupling of carbon dioxide is the model's prediction of a virtual cessation of the deep ocean salt transport current system. This result can be traced to the fact that a much warmer world

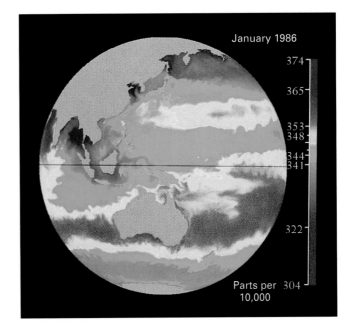

A "snapshot" of the amount of salt close to the surface in the oceans, modeled by Albert J. Semtner and Robert Chervin. Red represents areas of high salinity, blue of low salinity. The curves and swirls of eddies are particularly visible at the boundary between high and low salinity, in the tropics and south of Australia.

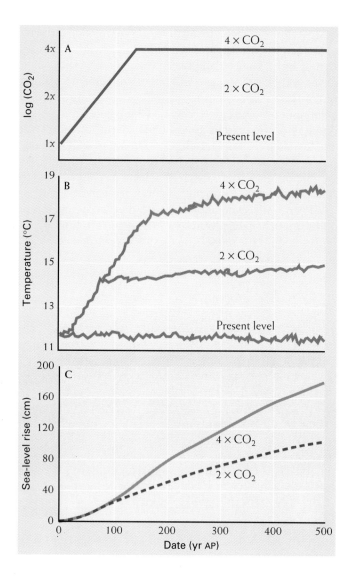

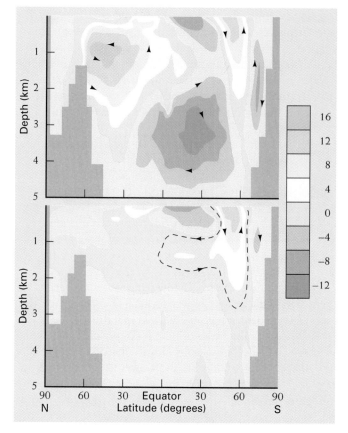

Results from the 3D global coupled ocean–atmosphere model of Syakuro Manabe of NOAA's Geophysical Fluid Dynamics Laboratory, designed to study the effects of varying the atmospheric level of CO_2. Results are shown for three computations: a scenario in which CO_2 levels are constant at today's level, one in which they are doubled in 70 years and then remain constant, and one in which they are quadrupled in 140 years and then remain constant. The left center and left bottom panels show the resulting global average temperatures and sea level increases. The right upper panel shows the present average latitudinal ocean currents (in units of 10^6 cubic meters per second and averaged over longitude) as represented in the model; the right lower panel shows the currents predicted by the calculation in which the CO_2 concentrations are quadrupled.

would evaporate more surface water from the continents and precipitate it over the oceans. In the northern hemisphere, where the majority of Earth's land mass is located, enhanced precipitation would cause a freshening of the North Atlantic and could result in a weakening or cessation of the sinking of the heavy saline water that drives

the salt transport system. In such a situation, which ocean-drilling data suggest has occurred a number of times in the geological past, heat would not be transported from ocean to ocean as it is today, and the ability of Earth to support life on the same scale and in the same locations as today would change dramatically—and not for the better.

Probabilities and Possibilities

Not all predictions for the climates and atmospheric chemistries of the future are thought to be equally probable, though all are worthy of study. To impose some sort of priority system on our thinking about these matters, Gerald Mahlman of NOAA's Geophysical Fluid Dynamics Laboratory suggests the following lists:

The Virtually Certain

The concentrations in the atmosphere of greenhouse gases will continue to increase, as a consequence of anthropogenic activities such as the burning of fossil fuels and expanding agriculture.

The radiative properties of increased concentrations of greenhouse gases will produce a net heating effect on the planet.

High-altitude cooling caused by the combination of reduced ozone concentrations and increased carbon dioxide concentrations will lower the upper-stratospheric temperatures by as much as 8 to 20°C, altering the atmosphere's circulation patterns.

Many centuries will pass before carbon dioxide concentrations will return to normal levels, even assuming that all anthropogenic emissions stop entirely.

The Probable

The global mean surface temperature will rise by 1 to 3°C by the middle of the twenty-first century.

The rate of water evaporation will increase as the climate warms, resulting in increases in global mean precipitation.

The mean sea level will rise 10 to 30 cm by 2050.

The Uncertain

A warmer, wetter atmosphere may increase the frequencies of tropical storms.

Increased precipitation in high northern latitudes may reduce the salinity and density of the ocean waters there, influencing global ocean circulation.

Mahlman's lists provide a perspective that may help us to make decisions in the midst of our uncertainty about the climate and atmosphere of the next few centuries. This question of how to respond constructively is worth exploring in some detail, and we defer it to the end of the book. Before that, in the next chapter, we attempt to look even further ahead than a few centuries. What does our crystal ball say about those more distant times, and what are the implications of its predictions?

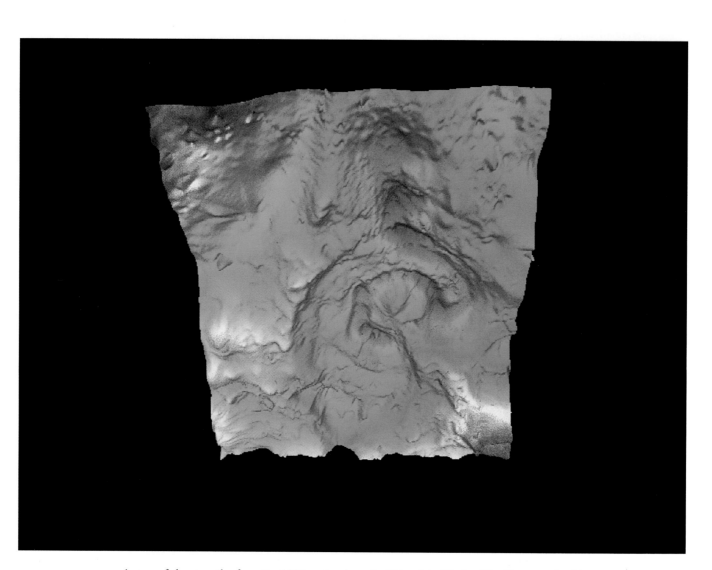

A map of the strength of gravity 1200 meters beneath Chiexulub, Mexico. The blue areas in this three-dimensional reconstruction represent regions where Earth's gravity is close to normal; colors grading from green to yellow to red represent regions with increasingly lower gravity than normal. Geophysicists believe that this pattern may be the remains of a crater left by dinosaur-killing impact of a bolide with Earth.

Predicting the Far Future

7

Forward, forward let us range;
Let the great world spin forever down
The ringing grooves of change.
—Alfred, Lord Tennyson,
Locksley Hall

Into the eternal darkness,
into fire, into ice.
—Dante, The Inferno

By this point, it should be clear that industrial and agricultural development, as well as rapid population increases accompanied by increased emissions to air, water, and soil, will greatly change Earth's environments over the next few centuries. This is a time scale that affects us personally: the changes that are accumulating now will accelerate during the lives of our children and grandchildren, and their grandchildren. But what is our stake in looking farther ahead than that, focusing on thousands or millions of years in the future? In considering that time scale—a scale on which the genus *Homo* can be expected to evolve into new forms—our concern for our own genus is outweighed by a more general concern for the life of our planet. On time scales of these magnitudes the natural forcing functions that determined the shape of Earth's geological history will assume greater prominence than anthropogenic ones, and a knowledge of the distant past gains importance as a guide to the future.

Neo-Cenozoic Climate

The Neo-Holocene ("new Holocene") is the term by which we designate the epoch extending as far into the future as the Holocene extends into the past: about 10,000 years or so. For that period, especially the early part of it, one factor will differ significantly from its condition in the Holocene epoch: the concentrations of greenhouse gases. As we have seen, those concentrations are now increasing relentlessly, and they seem likely to continue to do so for quite a while. Such projections are, of course, very uncertain, depending heavily on long-term patterns of fossil fuel use, the deforestation of the planet, and other factors, but concentrations several times those of the present are possible for carbon dioxide and other greenhouse gases. These extreme conditions could develop over a period of a thousand years or less, by which time the fossil fuel supplies of Earth could well be exhausted. Wallace Broecker of the Lamont-Doherty Geophysical Observatory in New Jersey proposes that the result of this large injection of greenhouse gases into the atmosphere would be a brief (on geological time scales) "anthropogenic super-interglacial" period, which is illustrated below.

In the Neo-Pleistocene epoch, extending from 0.01 to 2 Myr AP (after the present), one of the principal influences on climate is expected to be the variations in rotation and revolution of the Earth in its orbit. Just as

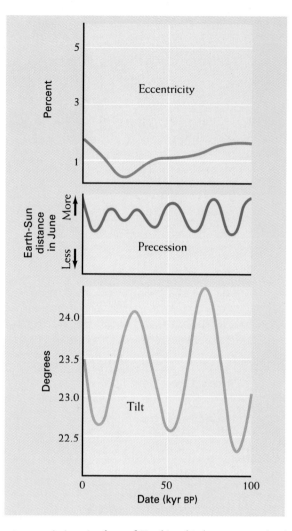

Long-term variations in three of Earth's orbital parameters: (top) eccentricity; (middle) precession; and (bottom) tilt, from the present time to 100 kyr AP.

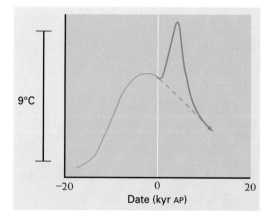

A schematic picture of the pattern of Earth's temperature during the Holocene epoch and its possible rise and fall during the first part of the Neo-Holocene. The dotted line indicates the pattern that would be expected in the absence of anthropogenic greenhouse gas forcing.

these variations have been calculated for the past, so too have they been calculated for the future, and it is interesting to compare the projections shown in the graphs above with the historical re-creation shown in Chapter 4 (page 77). Note first that the orientation of the rotational axis relative to the Sun–Earth line (the tilt) is roughly at its midpoint at present. For the next 8000 to 10,000 years, the tilt will decrease, thereby causing less solar radiation to reach each pole during its summer and increasing the chances for ice and snowpack buildup there. The eccentricity of the orbit of revolution is decreasing as well, a development that has a slight tendency to equalize seasonal variations. The axial precession (the measure of the Sun–Earth distance during a particular season) has the shortest period of the three parameters, and the magnitude of the distance is now decreasing in June and increasing in December; this too will encourage global cooling by reinforcing winter snow and ice cover.

Equations incorporating these changing orbital parameters were used to calculate the total radiation reaching the top of the atmosphere as a function of time, latitude, and season (summarized in the diagrams on this page). Because changes in radiation are most important in the northern high and midlatitudes, where the bulk of the land is presently found, and in the winter, when the reductions in radiation interact most directly with the maintenance of the snowline in the interior portions of the continents, the general decrease in irradiation in January as well as July at high northern latitudes for the next few tens of kiloyears is most relevant to our considerations. With the potential appearance of more glaciers at high latitudes, the enhanced reflectivity of radiation will encourage global cooling.

The relative impact on climate of variations in the orbital elements is demonstrated nicely by the study of monsoon-related climatic records illustrated below, in which the average northern hemisphere summer radiation flux during the past few tens of kiloyears is seen to vary, largely as a function of ellipticity and precession. In close concert with those variations, global average temperature (as measured by the oxygen isotope content of fossils), the occurrence of monsoon-related sediments, and the occurrence of monsoon-transported pollen all show four nearly coincident peaks.

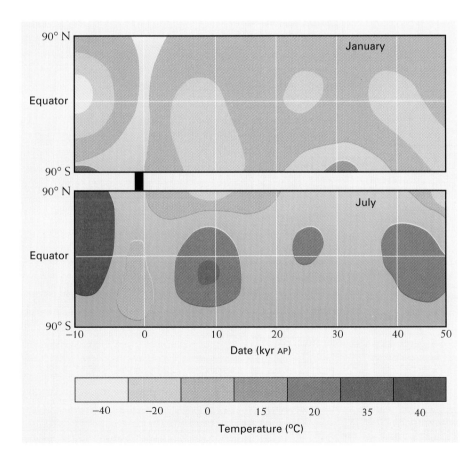

Changes from 10 kyr BP to 50 kyr AP in the amount of solar radiation reaching the top of Earth's atmosphere as a function of latitude, from pole to pole. (Top panel) January; (bottom panel) July. The units are cal cm^{-2} day^{-1}, expressed as departures from present values, which are of the order of 700 cal cm^{-2} day^{-1} and depend on latitude and season.

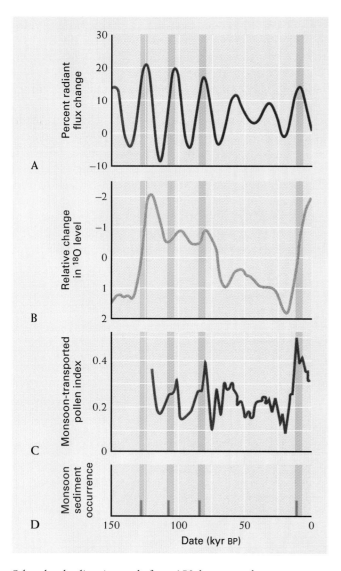

Selected paleoclimatic records from 150 kyr BP to the present as an illustration of the impact of orbital variations on indicators of climate. (A) Average northern hemisphere summer solar radiation, expressed as a percentage change from today's value; (B) composite temerature record (as reflected by fluctuations in the ^{18}O content of fossils); (C) monsoon-transported pollen in deep-sea sediments; (D) occurrence of organic-rich sediments associated with tropical African monsoon runoff.

Thus, the forecasts based on these orbital parameters, predicting that a global glaciation will gather force over the next few millennia, are supported by a variety of historical records. The temperature decrease would probably not be gradual, but would occur in irregular steps of about 1°C each. As we noted earlier, the difference between the temperature of the Little Ice Age from about AD 1400 to 1650 and that of the present is thought to have been about 1°C. Andre Berger of the Université Catholique de Louvain in Belgium has generated a series of predictions for climate over the next 100,000 years. Overall, they suggest that in the absence of anthropogenic influences, the long-term cooling trend that began some 6000 years ago will continue for the next 5000 years. This period of cooling is then expected to be followed by a cold interval centered at about 22 kyr AP and finally by a major glaciation culminating at about 50 to 60 kyr AP.

Predictions of glacial and interglacial cycles can in turn be used to predict continental ice volumes. Calculated in this way, the ice-volume pattern for the next 80,000 years is seen to follow the rise and fall of temperature rather closely, the principal difference being the asymmetry that results from the tendency for ice sheets to shrink faster than they grow and hence for major glaciations to end faster than they begin. The calculation (expressed as the red curve in the lower graph on the facing page) assumes that the Greenland ice sheet will disappear in the next few centuries as a result of greenhouse warming, a possible occurrence. In that case, the northern hemisphere ice sheets would not reappear before 15 kyr AP, and the climate would deviate from its predicted natural evolution until about 65 kyr AP.

An important but uncertain influence on Neo-Cenozoic climate is the degree to which the oceans will serve as a sink for carbon dioxide, thus buffering any warming influence that might also be operating at that time. A warmer ocean will dissolve less carbon dioxide, but that effect could be counteracted by the productivity of the marine biota, which serve as a biological carbon dioxide sink by transporting organic carbon from surface waters to the deeper ocean layers in a rain of detritus. The marine biota are thought to constitute 30% or so of the planet's biomass, thus reducing carbon dioxide concentrations substantially.

The latter part of the Neo-Cenozoic era, the Neo-Tertiary period, extending from 2 to 65 Myr AP, will be lengthy enough to witness changes in glaciation, aridity, and other temperature-related factors but too short to register tectonic movement of the continents, major changes in solar radiation intensity, or similar forcing processes oc-

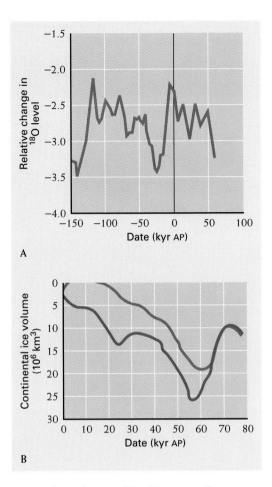

(A) *Long term future forecast of Earth's average climate, expressed as oxygen isotope ratios of deposited sediments.* (B) *Variations in continental ice volume for the next 80,000 years in the absence of ice sheet disturbance (purple curve) and if the Greenland ice sheet is assumed to melt over the next few centuries (red curve).*

tudes, may be reflected in climate records. A second example of climate change on such a time scale is the effect of bolide impacts. The timing and effects of bolide impacts have been deduced from the appearance and ages of craters on Earth and from calculations of variations in the orbital relationships of Earth, the Sun, and nearby planets and stars. If such impacts indeed follow the regular patterns that seem to be dictated by astronomical orbits

In the South American Andes, annual bands of ice growth are clearly visible in this glacier. The habitability of Earth's land areas is strongly influenced by the growth and dissolution of glaciers in response to changing climatic conditions.

curring over very long periods of time. Rather, climatic forces that cause changes on a time scale of 30- to 60-million-year intervals will be the important influences governing that period. For example, scientists think it takes about 35 million years to make progress in tectonic mountain building (as distinct from major continental relocation); and mountain building, which can increase glacier formation by adding to the amount of land at high alti-

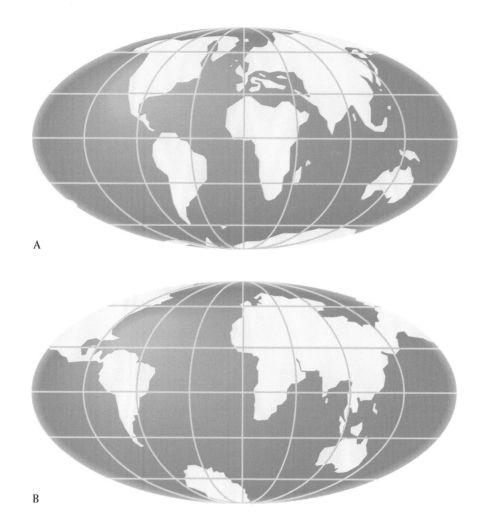

A

B

(see the figure on page 75), the last such encounter may have occurred about 15 Myr BP, and the next may occur about 11 to 15 Myr AP.

We noted earlier that the consequences of a bolide impact are expected to include rapid cooling and an inhibition of photosynthesis for a year or more, because solar radiation will be extinguished by nitrogen dioxide, soot particles, and perhaps suspended dust. In addition, highly acidic rain may cause near-extinction of the oceanic microorganisms whose shell making normally removes carbon dioxide from the atmosphere–ocean system, while, to compound that situation, the increased acidity of the oceans will decrease the solubility of carbon dioxide. The

quantities of dust injected into the atmosphere will deflect solar radiation and cause an initial period of cooling, but eventually the increasing atmospheric concentrations of carbon dioxide and other greenhouse gases will create a strong greenhouse warming that is likely to last thousands of years.

It is difficult to choose among these possible scenarios for the Neo-Tertiary period, and thus to forecast a likely trend to climate. If bolide impacts occur, they will encourage short-term cooling and long-time warming. If not, climate would be expected to follow its standard pattern of glacial and interglacial response to the variations in Earth's planetary orbit around the Sun.

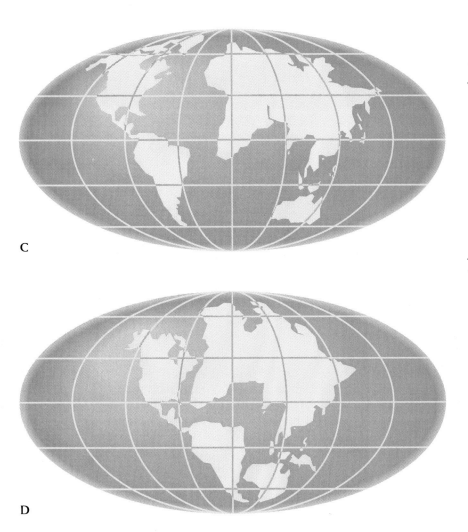

C

D

Current positions and prospective future locations of the continents. In the present condition (A), the American continents are joined, as are Africa and Eurasia, and India has collided with Eurasia to form the Himalayan mountains. (B) 100 Myr AP: Africa and Europe join completely, forming new mountain chains. Australia moves closer to Eurasia. The Atlantic Ocean exceeds the Pacific in size and the land bridge between North America and Asia strengthens. (C) 150 Myr AP: Eastern and western hemispheres separate as the Atlantic Ocean narrows. Land masses move toward the Equator. (D) 250 Myr AP: The closing of the Atlantic forces the continents toward the formation of a new supercontinent, containing a remnant ocean between Africa/Eurasia and South America/Australia/Antarctica. New mountain chains are formed by the continental collisions.

Neo-Mesozoic and Neo-Paleozoic Climate

The Neo-Mesozoic and Neo-Paleozoic eras extend over the approximate time range of 65 to 600 Myr AP. The significant forcing function for climate within this time span—one that also applies, in less significant ways, to shorter periods—is the alteration of land mass distribution caused by continental drift. Projections of changes in the dispositions of the continents are not guaranteed, of course, but the experts are in agreement concerning general patterns of movement.

Predictions for continental locations at 100 Myr AP are that the primary land masses will move farther north, while retaining much of their present arrangement and character. Some cooling can be expected, as occurs whenever the continents pass over the poles, but most of the climatic consequences of continental movements are likely to be local and regional rather than global.

Continental positions are expected to have changed significantly by 150 Myr AP. Plate motions by this time will have carried much of the continental mass toward the equator, decreasing the possibility of polar ice caps and making more of the land available to the equatorial Sun for the absorption of radiation. Thus, the tendency of continental drift during this period will be to increase the average temperature of the planet by decreasing its albedo. A supermonsoon situation is probable, with a very hot and wet climate over much of the continental area. These conditions are even more likely to prevail toward 250 Myr AP, when the continents may be nearly joined in a sprawling new supercontinent that extends over tropical and temperate latitudes. Little potential will remain under those circumstances for formation or retention of ice and snow, and thus planetary heating will be strongly reinforced.

Neo-Precambrian Climate

The farthest we can look into Earth's future is equal to the distance in time we can look into the past. On this time scale, a few billions of years, the major additional climate-forcing factor to consider is the aging of the Sun and its consequent effect on the intensity of solar radiation on Earth. Earlier we discussed the "weak sun paradox": the prevailing temperatures during the early life of the planet allowed liquid water to exist despite a solar radiation flux that is thought to have been much lower than at present. High levels of atmospheric carbon dioxide and a luminous, mass-shedding Sun have both been proposed in efforts to resolve the paradox. Astronomers' prediction for the far future is that the luminosity of the Sun will continue to increase, further warming Earth and its atmosphere.

The Neo-Precambrian perspective allows us to examine the ultimate question: Does life on Earth have a fixed time span, and can its end be predicted? To answer this question, we need to look at the prospects for three interrelated and vital features of the Earth system. The first is carbon dioxide, which is necessary for plant photosynthesis and thus for the maintenance of the biosphere as presently constructed. While plants vary in how much carbon dioxide they need to sustain the photosynthetic process, today's average carbon dioxide level of about 360 ppmv encourages a wide variety of plant species to grow. The minimum level, suitable for only a very small subset of plants, is thought to be about 10 ppmv. Should atmospheric carbon dioxide levels fall below that value, the biosphere's uptake of carbon, life's basic building block,

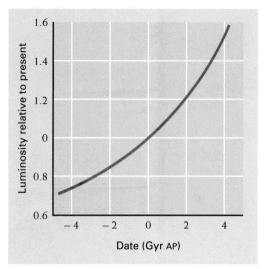

Total solar luminosity relative to that of the present from the time of the organization of the solar nebula and the assembly of the Sun and the planets at 4.7 Gyr BP until more than 5 Gyr AP.

would cease. The second crucial feature of the Earth system is temperature. Most life on Earth has evolved under reasonably stable temperature conditions, ranging from about 0 to 30°C. When average temperatures are greater than 30°C, higher forms of life begin to disappear, and above about 50°C only certain forms of bacteria are able to survive. The third critical feature is the presence of water. Not only is this unique molecule a part of every living cell, but it also mediates the cycles of dissolution and precipitation of the rocks that make up our continents. Water can be lost from Earth by being injected high into the atmosphere. There it is slowly broken down by high-energy solar radiation, and the resulting hydrogen escapes into space.

How will increased solar radiation influence these three ingredients so vital to life on Earth? The situation has been assessed in a computer model developed by Ken Caldiera and James Kasting of Pennsylvania State University. Their first finding, an obvious one, is that planetary temperatures will begin to rise. As this happens, temperature-dependent chemical equilibria will shift, making the soil increasingly acidic and causing Earth's abundant silicate rocks to weather more rapidly, releasing their calcium and magnesium ions into streams, rivers, and oceans. In the oceans, these ions will combine with dissolved carbon dioxide and precipitate as carbonate minerals. More atmospheric carbon dioxide will then dissolve to maintain its equilibrium between air and water, and thus the atmosphere's store of carbon dioxide is depleted.

The decrease in atmospheric carbon dioxide (graphed on the following page) will have two results: on the positive side, it will diminish the greenhouse effect and balance to some extent the increasing solar luminosity; on the negative side, it will make life increasingly difficult for plants. Caldiera and Kasting predict that carbon dioxide levels might drop below 10 ppmv by about 1 Gyr AP.

Even without the carbon dioxide to warm the atmosphere, planetary temperatures will rise, due to the increase in solar luminosity. At about 1.3 Gyr AP, the average temperature may exceed 50°C; at 1.5 Gyr AP, it could reach 100°C. The continuing rise in temperature will increase the rate of water evaporation from lakes, rivers, and oceans—and hence the gradual loss of water from Earth.

The rate at which substantial water depletion will occur is not certain, but Caldiera and Kasting's calculations imply the disappearance of photosynthetic plants at about 1 Gyr AP, the extinction of life forms other than bacteria at about 1.3 Gyr AP, and the end of all life on Earth between 1.5 and 2.5 Gyr AP.

A Summary of Possible Climate Futures

A quick review of the long-range climate predictions described in this chapter begins with the Neo-Holocene epoch, when increases in the concentrations of greenhouse gases, in conjunction with a related decrease in Earth's albedo as ice and snow areas are decreased, will most likely be more than sufficient to overcome the cooling of Earth as a result of its natural oscillatory cycle for the next thousand or so years. If this overcompensation does occur, Earth's temperature during the Neo-Holocene

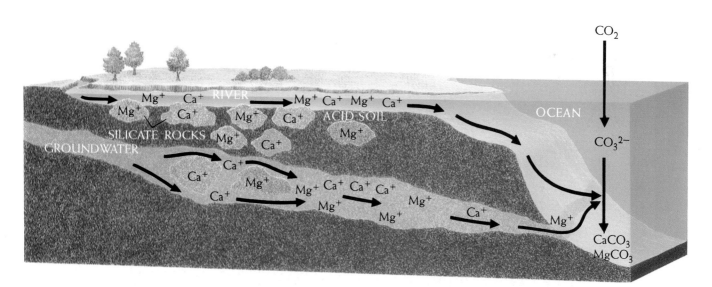

The fate of calcium and magnesium ions in silicate rocks under a scenario of increased global temperature.

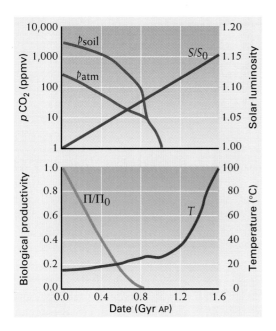

Computer model predictions for the Neo-Precambrian era of solar luminosity (S/S_0), temperature (T), atmospheric and soil CO_2 (P_{atm}, P_{soil}), and biological productivity (Π/Π_0). Below about 10 ppmv CO_2 plants are unable to survive; this situation is predicted to occur shortly before 1 Gyr AP.

A swollen Sun—half the size of the orbit of Mercury—rises as a red giant to blister Earth.

epoch will exceed its highest level in more than 100,000 years. Following that warm spike, however, the climate will resume its cooling trend for the ensuing several thousand years, reaching the next ice age at about 5 kyr AP. After that, natural cycles of interglacial and glacial periods will continue.

For the Neo-Tertiary period, the possibility of bolide impacts must be figured into the climatic equation, along with the forces of greenhouse gases and the planet's orbital cycles. For a few years following a bolide impact, Earth would probably be cooler, and substantial extinctions of many of its life forms could occur. Over a longer term, however, the additional carbon dioxide and other greenhouse gases that the impact would release into the atmosphere would provide increased impetus for global heating. Major impacts on the biosphere, including *Homo sapiens*, are clearly possible. The probability of a bolide impact as sizable as the one that apparently occurred at the Cretaceous–Tertiary boundary, or so recently on Jupiter, is rather small during the lifetime of our genus (perhaps 10 million years). However, there is a much greater probability that smaller but still huge impacts—with energies many times larger than that of a full nuclear war—will occur before the descendants of *Homo sapiens* have walked their last upon Earth.

In the Neo-Cenozoic and Neo-Paleozoic eras, the movement of the continents toward the equator is expected to bring an additional and increasingly important heating process into play: most of the land masses will be in positions conducive to absorbing radiation and *not* conducive to retaining highly reflective ice and snow. It is quite uncertain what temperatures can be anticipated when this happens, but they may well be hostile to many of the planet's current animal and plant species. Evolution is a resourceful and powerful phenomenon, however, and the ability of some living things to adapt to extreme conditions is demonstrated by the abundant life found today at high temperatures and high pressures near the hydrothermal vent sites on the ocean floor. Neo-Cenozoic and Neo-Paleozoic life may therefore differ from the forms we find familiar, but it would be naive to assume that no life of any kind will be present.

Over the longest future time span of geological interest, the Neo-Precambrian era, the aging of the Sun will produce a large increase in solar flux, causing radiation intensities on Earth to increase by 50% or more. Carbon dioxide will diminish under these circumstances, and photosynthesis will cease. Earth will become warmer and warmer as the Sun proceeds along the path of its demise. Eventually the oceans will begin to evaporate, and gradually the water vapor will lose its hydrogen to space. In time, perhaps by 2 to 3 Gyr AP, Earth's miracle fluid will have been lost completely—and with it the planet's remaining inhabitants. At that point, the only hope for the continuance of Earth-engendered life may be the construction of mini-biospheres elsewhere in space.

The Sun rises over waves of grain from Nebraska soil rendered fertile and productive by favorable weather and climate conditions.

Of Change and Sustainability

8

Upon the whole I am much disposed to like the world as I find it, and to doubt my own judgment as to what would mend it. I see so much wisdom in what I understand of its creation and government, that I suspect equal wisdom may be in what I do not understand.
—Benjamin Franklin

*T*oday's technological society encourages us to think that we have the capability to solve all our problems, perhaps including those of global change. The easiest way to do so, of course, is to deny that the problems exist in the first place, as some have tried to do by challenging the concepts and ideas put forth by the scientific community. Scientists, in turn, must refute the arguments of the skeptics. There are also some Earth scientists who hope to counter the worrisome global changes we have been discussing through heroic high-tech actions, others by implementing multitudinous small changes in today's technology. There is even a small minority who believe Earth can cure itself.

155

Throughout this book we have focused on the idea that the environments of Earth, always subject to change, have in recent times been changing at least as rapidly as at any other point in history—with the exception of transient catastrophic events such as volcanic eruptions and bolide impacts. Plotting the time scales of various historic events, as we have done in the diagram to the right, helps put some of those changes in perspective. The ordinate on our diagram is logarithmic, so each step up from the bottom represents a time span 10 times the length of the segment just below it. At the top are the factors whose influences operate over the longest time scales: the evolution of the Sun into a red giant star and the alterations in the location and morphology of continents produced by plate-tectonic motions. The interactions of both with climate and chemistry are indisputable, but they occur over such vast stretches of time that humankind will not be affected materially by them.

At the next level are major bolide impacts. In Chapter 7, we saw that these have a typical frequency, as suggested by both the geological and fossil records, of perhaps 25 to 30 million years. These impacts are natural events with immediate and devastating effects. Less catastrophic, but still drastic enough to completely alter the physical and biological character of the planet, are the ice ages, whose cycle times of 10,000 to 20,000 years are not much longer than the written and artifactual record of human civilization.

The atmospheric emissions attributable to humankind exert their climatic effects on three different spatial scales: global, regional, and local. Overall, the effect of our industrial and agricultural activities has been to introduce a surge in the spectrum of Earth's climatic variability on a time scale of about 100 years. Only intermittent natural catastrophes, such as very large volcanic eruptions and bolide impacts, have a more immediate effect on the atmosphere.

Compare the various scales of climatic change with some of the typical lifetimes of organisms and features on the planet. One striking coincidence is that the release of anthropogenic emissions into the atmosphere is causing global change over a time span consistent with typical human life spans. As a consequence, both the fruits and the repercussions of our actions will be visible to us and to our

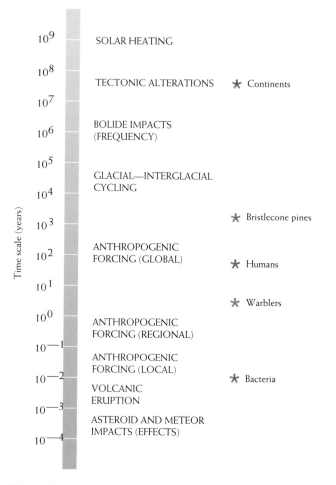

Time scales on which occur the events responsible for atmospheric change. Typical lifetimes of several terrestrial features and inhabitants are noted at the right for comparison.

children, our grandchildren, and their grandchildren—not delayed until some time too distant to cause us personal concern.

Two Cheers for Predictions

Making models of the Earth system is one of the best ways available to improve our understanding of complicated problems in Earth system science and predict the future impacts and consequences of our actions. However, we must keep three things in mind about such models. First, computer models of the Earth system are very detailed and complicated, and it is not always possible to tell whether all the factors and formulations they are based on are perfectly correct. Second, two similar, well-constructed computer models can produce significantly different results. Third, notwithstanding such inconsistencies, the results from the models are useful tools for improving out understanding of the Earth system of today and for predicting the chemistries and climates of the future. If we cannot, at their present state of reliability, give models the traditional "three cheers," we can perhaps give them two. At the very least, any disturbing predictions made by models should be treated as important warning signals.

The characteristics and capabilities of models are regarded differently by different people. Generally, however, a scientist sees the results of a model not as doctrinaire predictions, but as guides to understanding and as pointers to areas needing further study. Models, laboratory experiments, and field measurements all interact to increase our knowledge, as illustrated by the various efforts over the past decade to understand the causes and consequences of the ozone hole. The process began with the initial discovery of the ozone hole in an extended series of field data, as discussed in Chapter 5. Models were constructed to explain this startling finding, but none was able to reproduce the hole until Susan Solomon and her colleagues at the National Oceanic and Atmospheric Administration in Boulder, Colorado, included polar ice particle effects. In their models, the particles accelerate the conversion of hydrochloric acid and chlorine nitrate ($ClONO_2$), which do not normally react with ozone, into the "ozone killers" Cl· and ClO·. Guided by these model results, detailed laboratory studies of reactions on frozen particle surfaces were undertaken and indeed showed that the conversion rate is markedly enhanced in the presence of particles of ice. Models then predicted that large concentrations of ClO· would be seen where ozone was depleted, and aircraft and satellite observations confirmed this. It was next discovered that the particles consisted in most cases not of water ice, but of a 1:3 molecular mixture of nitric acid to water. These solid particles can be formed at

a warmer temperature 10°C than that needed to form water ice particles, thus strongly enhancing the efficiency of formation of the "ozone killers." Although some interesting scientific questions remain to be addressed, the interaction of field observations, laboratory studies, and computer models has produced in only about five years a virtually complete understanding of this recently recognized, complicated problem in atmospheric chemistry.

At all stages of this investigative process, the theory of ozone depletion has been under attack. While those who are not intimately involved in its study can be forgiven for asking whether the CFC–ozone theory is really a hoax, they are raised so often, that it behooves us to address the most important ones here.

▶ **The existence of the ozone hole is merely speculation.**

Data from several sources, including the ground-based and satellite observations described in Chapter 5, counter this statement. Furthermore, the measurements show that since its discovery in 1985, the ozone hole has deepened further. In the austral spring of 1993, the concentrations of ozone detected over Antarctica registered new all-time lows.

▶ **CFC molecules are heavier than air and will never reach the upper atmosphere.**

The atmosphere is not a calm blanket of air where gravity reigns supreme, but a turbulent fluid in which light and heavy gases are mixed at equal rates by the motions of large air masses. Among the numerous field observations that demonstrate the effects of this mixing are those carried out by balloon-borne measuring instruments, described in Chapter 3, and the CFC-12 observations from the Upper Atmospheric Research Satellite (UARS) diagrammed on the next page, all clearly showing that CFCs are transported high into the stratosphere.

▶ **Volcanoes put far more chlorine into the atmosphere than CFCs do, so the latter are unimportant.**

The occasional large volcano can indeed release substantial amounts of chlorine as hydrochloric acid into the at-

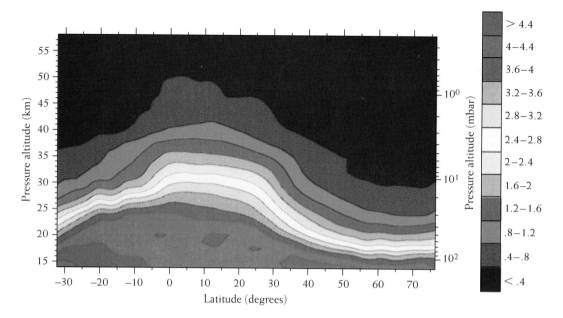

A two dimensional depiction of measurements taken by the Upper Atmosphere Research Satellite, showing CFC-12 concentrations throughout the stratosphere. Higher concentrations in the northern tropics and midlatitudes reflect the higher rates of emissions there. Nonetheless, CFC-12 is seen to spread from those sources to high altitudes and far-distant latitudes.

mosphere, but hydrochloric acid, unlike the CFCs, is very soluble in water and therefore most of it is rained out in the lower atmosphere. Stratospheric hydrochloric acid has been monitored at Kitt Peak National Solar Observatory since 1977. Data gathered there show that even after the 1982 eruption of El Chichón, in Mexico, probably the most significant injection of chlorine into the stratosphere ever observed by scientists, no clear enhancement of hydrochloric acid occurred. Moreover, the ratio of hydrofluoric acid to hydrochloric acid in the stratosphere remained generally stable, equal to the ratio observed when both fluorine and chlorine have come from CFCs.

It is often claimed that the Mt. Erebus volcano in Antarctica releases large quantities of chlorine into the stratosphere. These releases have been measured, however, and prove to have chlorine emission rates less than 1% of that resulting from global CFC production. Furthermore, Mt. Erebus does not produce explosive eruptions, so its

emissions are carried off into the troposphere by low-level winds and subsequently washed from the atmosphere.

▶ **Sea salt emissions are much larger than CFC emissions.**

The amount of chlorine in particles from sea spray is indeed large, but sea salt particles are rapidly returned to the surface by gravitation and by precipitation. Almost none reaches the stratosphere, as shown experimentally by the negligible amounts of sodium found there.

▶ **The ozone loss is part of a natural cycle and has nothing to do with CFCs.**

Because the proposed ozone loss reactions involve the molecular fragments Cl· and ClO·, the relationships of these constituents can be tested by observing what happens to ozone levels when Cl· or ClO· concentrations rise or fall. The ClO· radical is easier to measure, and it is now being tracked by optical detection from ground stations, by aircraft-borne instruments, and by detectors on orbiting satellites. The figure on page 160 shows ClO· and ozone maps based on data collected by the UARS satellite during the annual ozone cycle decrease. On these maps, the strong ozone losses are seen to have perfect spatial coincidence with enhanced concentrations of ClO·.

Nonscientists often equate the word "theory"—as in the theory of ozone loss or any other scientific theory—with speculation. As scientists use the term, however, a theory rests on much more solid ground; it is an attempt to incorporate a number of facts into an explanatory picture of how nature works. The facts themselves are indisputable, provided that they result from work that has been substantiated through careful replication: the atomic weight of chlorine is known, the rate of reaction between ozone and chlorine atoms has been measured, and so forth. The picture constructed from these facts—the theory—is a conceptual framework to be probed and tested for structural faults and instabilities, and this probing and testing for the strengths and weaknesses of each theory is how science advances. In the case of the ozone hole, the pro-

bing and testing have produced a level of understanding as sophisticated and well-supported as that in virtually any other area of science.

A second Earth system theory that has been subjected to extensive scientific and media discussion is that of global warming. In contrast to the explanation for the ozone hole, a theory that has largely been confirmed by experiment, the global warming theory is still immersed in basic experimentation, refinement of explanatory models, and a search for the critical processes. Global warming is a much less tractable issue than that of ozone loss in several ways: it involves not only the atmosphere but also the oceans and terrestrial vegetation, so it is very much more complicated; the time period required for detecting a significant warming effect and for correcting it is of the order of centuries, not years or decades; and any solution to the problem would necessitate major economic investments and lifestyle changes.

Two central tenets of the global warming theory are based squarely on experimental fact and are agreed on by all participants in the debate. The first is that the greenhouse effect is currently in operation on Earth and has operated during virtually the entire life of the planet (see Chapter 2). The absorption of radiation by gases in the atmosphere (chiefly carbon dioxide and water) raises

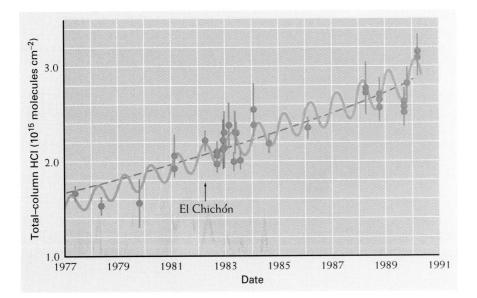

HCl total vertical column concentrations derived from observations at the Kitt Peak National Solar Observatory, Arizona, for the period 1977 to 1991. The solid curve is a fit to the data assuming that a linear increase with time (the dashed line) is superimposed on a sinusoidal seasonal cycle produced by atmospheric transport and dynamics. The date of eruption of the El Chichón volcano is indicated.

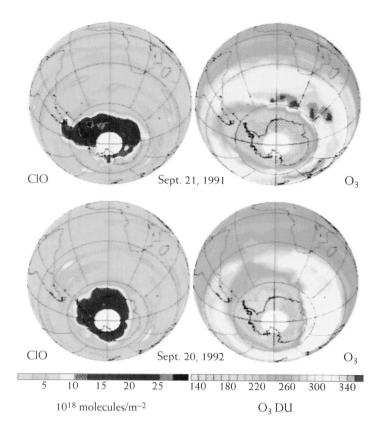

CIO   Sept. 21, 1991 O_3

CIO   Sept. 20, 1992 O_3

5   10   15   20   25   140   180   220   260   300   340

10^{18} molecules/m^{-2} O_3 DU

Upper Atmospheric Research Satellite maps of chlorine monoxide (left) *and ozone* (right) *on September 20, 1992* (bottom) *and September 21, 1992* (top). *The correlation between enhanced C10• and reduced O_3 readily evident, as is the significantly lower ozone concentration in 1992.*

Earth's average temperature above the freezing point of water and so permits life as we know it to exist. The second accepted fact is that the atmospheric concentrations of a number of greenhouse gases are increasing due primarily to the actions of humans. We illustrated these increases in Chapter 4 for carbon dioxide and methane; similar data have been presented elsewhere for nitrous oxide, CFCs, and other gases.

What objections are raised to the theory of global warming? Are they "infrared herrings," to use climatologist Stephen Schneider's turn of phrase? Here are the most prominent arguments against the global warming predictions, with some of the counterarguments that Earth system scientists have used to refute them.

▶ **Earth's temperature fluctuates naturally, so temperature change is a natural process.**

This statement is completely true, as was shown in the final figure of Chapter 4. However, the concern about global warming can be likened to a concern one might have about attaching a weight to a spring without knowing the spring's strength. Will the spring retain its shape, and function properly, or will the weight distort the spring permanently? The radiation-absorbing greenhouse gases perform a function in the climate system similar in some ways to that of a weight on a spring. Any added weight will affect the oscillations of the spring, although if the weight is very small, those changes might not be noticeable. As the weight is increased, however, the effects become obvious, and when the weight is large enough, the spring distorts or breaks.

Greenhouse gas concentrations are obviously not high enough to cause major distortions in climate, at least not yet, but it is unclear whether or not the distortions are presently so small as to be unmeasurable. In any case, the planet is now about as warm as it has been in at least the past 150,000 years; further warming would put us into new territory so far as the tenancy of *Homo sapiens* on "Spaceship Earth" is concerned.

▶ **New analyses of data show that temperatures in the United States have not increased in the past century.**

Global warming data are based on planetary averages, not the record of 5% of Earth's surface; those averages show a clear global temperature increase of about 0.6°C during this century, with some regions experiencing small decreases, some little change, and the majority modest increases.

▶ **The warming of the past century is due to solar influences, not greenhouse gases.**

This idea is based not on any theory of how solar activity would influence planetary temperature but on a possibly coincidental similarity between changes in sunspot number (an indirect indication of solar radiation) and global Earth temperatures. Orbiting satellites, however, have made direct measurements of the Sun's radiation since 1980; they find such small variations that any long-term temperature changes caused by solar radiation variation could not possibly explain the observed increase.

▶ **Global warming will be beneficial.**

This position neither supports nor denies global warming theory, but asserts that if Earth does heat up, both the higher temperatures and the increased carbon dioxide will accelerate plant growth and improve life on the planet. This last assertion is a topic of vigorous scientific debate. Whatever amount of truth it might contain, however, it is hard to imagine that residents of the world's lowest-lying shorelines and of the various island nations, especially in highly populated Asia, would view a sea-level rise of 0.5 to 1.5 m in less than a century as a benefit while they watch their homes being flooded away.

A second problem with this proposal is that its promise of enhanced vegetation applies primarily to managed

A cartoonist's view of global warming, suggesting the perplexity of the average lay person regarding the issue.

crops, not to wild flora that have evolved naturally. Plants growing in nature are not usually able to move quickly from the geographical regions in which they have been flourishing into newly apposite regions created by significant climate changes. Thus, the likely result of any climate variation more rapid than climate variations that occurred naturally over the past 10,000 years, during which most of our native plants evolved, would be the irretrievable loss of many species of native vegetation. As the vegetation goes, of course, so go the insects, birds, and animals that depend on it for food and shelter. Rapid change of climate is, therefore, an invitation to major loss of species diversity.

Perhaps the most important question concerning global warming is one that cannot be addressed scientifically by models (at least not at present) and that is easily scorned by the global warming skeptics: How likely are we to push global climate out of its present relatively stable state to some radically different and far less hospitable one? To use the stability analogue of Chapter 1, instead of pushing the state of the climate a bit further up the hill, will we push it over a precipice? That prospect *is* a real possibility. To begin with, it is obvious that most of Earth science's predictions concerning the state of the global atmosphere over the next few centuries are not encouraging. The very long lifetimes of many of the chemicals being emitted into the atmosphere mean that long-term change, once begun, will be difficult to halt. Thus, although we are uncertain about the absolute accuracy of even the most sophisticated global model predictions, the preponderance of evidence makes it difficult to avoid the conclusion that the atmosphere will change noticeably over the next century no matter what steps are now taken—and that the effects on Earth will be highly significant.

It is also the case that the complexity of atmospheric chemistry suggests that a single action can have both positive and negative effects on the environment. A good example of this is the possible enhancement of atmospheric HO·, a radical that is vital as an atmospheric detergent. Its concentrations can be decreased by added methane and carbon monoxide or increased by added nitric oxide (which will, however, increase toxic tropospheric ozone). Methane and (to a minor extent) carbon monoxide are greenhouse gases, however, so their influence is not restricted to impacts on HO· concentrations. In fact, a number of gases influence not only HO· but also such properties as global temperatures, tropospheric and stratospheric ozone, the acidity of precipitation, and a host of other atmospheric attibutes. Hence, a decision intended to optimize HO· concentrations can clearly not be made without considering all the many ramifications. We remain largely in the dark about the details of cause and effect for rapid climate transitions, yet we know many have oc-

curred in the past. Thus, the development of a solidly grounded approach to global change, based on an overall assessment of all positive and negative effects, and receiving support from policymakers worldwide, is one of the great challenges humanity will be faced with in the future.

Countermeasures and Consequences

In recent years several proposals have been made for counteracting some of the negative environmental consequences of human activity, including global warming, stratospheric ozone depletion, and photochemical smog. Many of the suggestions are drastic, prescribing major manipulations of the environment. Should we seriously consider acting on these ideas, in the same way that we consider building sea walls or dredging entrances to harbors, or are they foolhardy notions that could all too easily turn out to be "cures worse than the diseases"?

Countermeasures to Stratospheric Ozone Depletion

Earlier we noted that the severe ozone depletion detected each spring over Antarctica is a consequence of reactions taking place on polar stratospheric cloud particles. These reactions convert the chlorine atoms sequestered in hydrochloric acid (HCl) and chlorine nitrate ($ClONO_2$) into the chlorine-destroying radicals Cl· and ClO·. To counter this process, Ralph Cicerone and Scott Elliott of the University of California, Irvine, and Richard Turco of the University of California, Los Angeles, proposed in 1991 that it might be effective to inject ethane (C_2H_6) or propane (C_3H_8) into the Antarctic stratosphere. They reasoned that the reactive chlorine radicals would react rapidly with either of those hydrocarbons, and become sequestered in HCl:

$$Cl\cdot + C_2H_6 \text{ or } C_3H_8 \rightarrow$$
$$HCl + C_2H_5\cdot \text{ or } C_3H_7\cdot \qquad (8.1)$$

To evaluate their hypothesis, the researchers carried out computer model calculations that simulated ozone hole development both in the absence and presence of added

C_2H_6 or C_3H_8. The model suggested that the ozone hole, which generally begins to form around the end of August of each year, would be suppressed substantially by sufficient hydrocarbon additions, as seen in the figure on this page.

Why was this proposal not adopted? For one thing, a fleet of airplanes was not conveniently available for the purpose. To deliver the 50,000 tons of hydrocarbons that would be needed annually, several hundred large airplanes would be required and their emissions would need to be considered, as would the logistical requirements of generating and transporting the hydrocarbon gases to the continent most distant from the developed world. Balloons might be used instead of airplanes, but would present logistical difficulties at least as great.

Cicerone, Elliott, and Turco are highly respected atmospheric chemists, and their work stimulated considerable thought, if little action. That lack of action was proved fortunate in 1994, when the authors repeated their calculations with what at first seemed a minor variation, the addition of two reactions, including:

$$HCl + HOCl \rightarrow H_2O + Cl_2 \qquad (8.2)$$

Previously neglected in most atmospheric chemical models, the connection to the atmospheric chemistry of the hydrocarbons is that hypochlorous acid (HOCl) production is enhanced via

$$ClO\cdot + HO_2\cdot \rightarrow HOCl + O_2 \qquad (8.3)$$

as hydrocarbon breakdown generates substantial amounts of $HO_2\cdot$. When the appropriate reaction velocities and chemical concentrations were included in the model, the addition of hydrocarbons to the Antarctic stratosphere was seen *not* to lock up reactive chlorine as HCl, but rather to enhance its conversion to Cl_2. Recall from the earlier discussion of the ozone hole that Cl_2 readily vaporizes from polar cloud particles and then dissociates upon encountering sunlight, forming the ozone-killing $Cl\cdot$ radical:

$$Cl_2 + sunlight \rightarrow 2\ Cl\cdot \qquad (8.4)$$

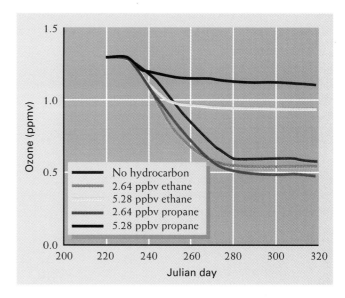

The development over time of ozone concentrations at 15 km altitude for several models of Antarctic stratospheric chemistry in which inorganic chlorine was set to 1987 levels and different amounts of ethane or propane were injected. (Julian day 1 of each year is January 1.)

Hence, the results from this second research project completely upset those of the first. The addition of ethane or propane to the Antarctic stratosphere was found to enhance ozone loss, not diminish it.

Countermeasures to Greenhouse Warming and the Next Ice Age

Greenhouse warming arises from the release into the atmosphere of gases that absorb the infrared radiation emitted by Earth, thus preventing the escape of the radiation into space. To counter this warming, some experts have proposed steps intended to reduce the amount of radiation reaching Earth from the Sun in the first place. The same idea might also be proposed for solving the more distant problems that will follow from the long-term increase in solar luminosity as the Sun ages. If effective, this latter countermeasure would result in a "strong Sun paradox," in

Conceptual diagram for solar reflectors placed at Lagrangian point L1 of the Earth–Sun system to decrease the amount of solar radiation incident upon Earth. There are five Lagrangian points, three of which are shown (the other two are at the points of equilateral triangles from the Sun–Earth line). Objects located at these points are stable, because gravitational acceleration and centrifugal acceleration are exactly in balance. L1 is approximately 1% of the distance from Earth to the Sun.

which the planet might be maintained at temperatures suitable for life even as solar radiation increased markedly.

What can humans do to diminish the amount of solar radiation received on Earth? One possible technique might be to send mirrors by space satellite to the Lagrangian L1 point (a point along the Earth–Sun line where no net forces act on a small object), so that the mirrors might be able to remain in place and reflect radiation indefinitely. Unfortunately for the concept, gravitational displacement forces from planets and other celestial objects are to be expected at the L1 point, so the mirrors would require an active positioning system to remain in place. Active positioning systems imply such things as mechanical components and compressed gases, and preliminary assessments of this idea suggest that spacecraft stabilization over long periods of time may be impractical. In addition, of course, the enterprise would be extremely expensive, not to mention beset with political intricacies.

An alternative to the reflecting mirror idea was proposed some years ago by Russian climatologist Mikhail Budyko, who envisioned injecting some 35 million tons of sulfur dioxide annually, about 25% of the amount presently released by fossil fuel burning, directly into the stratosphere. He calculated that such an amount, once converted there to sulfate aerosol, should significantly enhance the backscattering of solar radiation to space and cause a cooling of Earth's surface. A potentially more

practical variation on this suggestion is to inject soot particles rather than sulfur dioxide. The soot would not reflect solar radiation but would absorb it instead, and so efficiently that only a few percent as much soot would be needed to intercept as much radiation as all of Budyko's sulfate particles. In addition, the heat absorbed in the lower stratosphere at high latitudes might be sufficient to prevent the development of the polar stratospheric clouds and the ozone hole.

What are the prospects for any of these ideas working as intended? We discussed earlier the current effects of anthropogenic sulfur gas emissions, which form aerosols that appear to have counteracted some of the heating caused by elevated carbon dioxide concentrations. This finding suggests that, if properly executed, the ideas of Budyko and others might achieve their goals. However, the logistical difficulties of delivering thousands of tons of particles or gases are immense—not to mention the price tags. Whether the particles are loaded into ballistic shells and shot into the stratosphere by the world's large naval vessels (several thousand rounds per day, day after day, year after year), or whether they are transported to the proper altitude by fleets of hundreds of thousands of planes, the cost would be in the tens of billions of United States dollars annually.

Ironically, scientists are already thinking about proposals for improving the global climate many thousands of years from now, when the goal will probably be to warm the planet rather than cool it down. At that time, natural cycles are expected to be moving toward a new ice age, and greenhouse gas warming of the atmosphere will have diminished as fuel for greenhouse gas emissions becomes scarce. One proposed solution is to manufacture and inject greenhouse gases with highly efficient absorption properties into the atmosphere. A possible gas of choice is carbon tetrafluoride (CF_4), which is virtually unreactive and insoluble in water and thus would be expected to have an atmospheric lifetime of thousands of years. Its strongest absorption of radiation is centered in the infrared window, at a wavelength of about 8 μm. Thus, carbon tetrafluoride has all the properties of a powerful greenhouse gas.

A Prescription for Bolides

A problem very different from the ones described above is whether it is possible to lessen the consequences, apparently realized numerous times over the eons, of an asteroid or comet colliding with Earth. Such a collision, if large and relatively direct, would render habitation of the planet difficult or impossible for many of its biological species, including humans. Our current bolide-surveillance methods are not adequate to detect such objects routinely; nor have we any means of preventing a collision were the orbital paths of Earth and an approaching bolide discovered to be aligned. However, the necessary observation and intercept technology is not qualitatively different from that used to detect and intercept intercontinental ballistic missiles, and the past several decades have witnessed major progress in both those capabilities. Of course, if scientists were to continue and enhance antiballistic missile research and countermeasure activities to safeguard the planet from bolide impacts, it would be necessary to ensure that the resulting hardware was employed for peaceful purposes only. Though unlikely, it is not inconceivable that the world's military hardware could be turned from defense of national boundaries to defense against attack of Earth by random motion of celestial objects.

It is important to realize that several of the environmental countermeasures described above share three daunting characteristics: they assume a near-perfect knowledge of the effects that would be produced were they to be carried out, they require major logistical efforts, and they are expensive. Furthermore, they all depend on the peoples of Earth being able to continue to implement them indefinitely, regardless of changing political situations, economic conditions, and degrees of cooperation among nations. These are possibilities that only the very bold would be willing to bet on. Nonetheless, one could imagine an agreement on the logistics, economics, and oversight issues, leaving only the matter of perfect knowledge. It is here that all these ideas founder. To paraphrase a comment by John Firor of the U.S. National Center for Atmospheric Research, if we do not understand how a leaf works, how can we think we can make the atmosphere dance to our tune? A total understanding of the Earth system appears so unlikely that environmental countermeasures such as those discussed here, in our opinion, should not be considered—except in the face of very grave and unremitting disasters of our own making, and then with the utmost caution and flexibility.

Surprises

Most discussions of future possibilities begin at or near the present time, anticipate moderate changes in important environmental properties, and predict in some analytic way the consequences of these perturbations. Such techniques may appear to be quite satisfactory for most perturbations on short time scales. For the longer time scales that we have discussed in this book, however, they may be less helpful. All kinds of history, including particularly the history of climate, have shown that on time scales of several decades or more the important changes are generally not the result of gradual modifications of known factors, such as the price of oil or the global average temperature, but instead occur as a consequence of rapid and dramatic surprises. C. S. Holling, currently of the University of Florida, has defined such surprises as "when causes turn out to be sharply different than was conceived, when behaviors are profoundly unexpected, and when action produces a result opposite to that intended—in short, when perceived reality departs *qualitatively* from expectation."

Surprises can result from the presence of thresholds or nonlinearities that scientists did not know about and hence could not take into account in their calculations. A classic example, cited by Harvey Brooks of Harvard University, is the phenomenon of gridlock in urban areas at peak traffic hours, caused by individuals who drive automobiles in order to maximize their flexibility and efficiency in organizing working, shopping, recreation, and other commitments. However, too many such participants acting for their own benefit precipitate trouble for all, because the negative consequences of their actions set in quite abruptly at a critical traffic density, and their lives, instead of being improved, deteriorate more than if driving were abandoned altogether. Heavy automobile traffic also provides an example of a nonlinear *atmospheric* response: the generation of ozone from automotive emissions. Because the ozone is an indirect consequence of reactions among several emittants and because smog products from the reactions increase out of all proportion to the concentrations of emittants, high levels of urban ozone become virtually permanent at a certain critical level of urban development unless emissions from each automobile are reduced to extremely low levels.

As scientists learn more about the history of the Earth system, more and more of these surprises are coming to light. Among numerous examples we could cite are the mass extinction of the dinosaurs at about 65 Myr BP and the so-called punctuated (episodic, as opposed to gradual) evolution of animal and plant species during several other geological periods. We have already mentioned the unexpected—and sudden—decreases in stratospheric ozone over the Antarctic in the austral spring. Another intriguing example of the episodic nature of many Earth system phenomena is found in the work on ocean circulation done by Wallace Broecker of the Lamont-Doherty Geophysical Observatory. Recall that, in rough overview, the major subsurface ocean currents are initiated by the downwelling of cold, highly saline surface water in the North Atlantic Ocean, especially below the sea ice that is formed during winter. The subsequent flow of this water into the Pacific Ocean is sustained by the temperature differences between the two oceans, the higher temperature of the North Atlantic (relative to the cold, dry air there) causing the water to evaporate (thus promoting high salinity

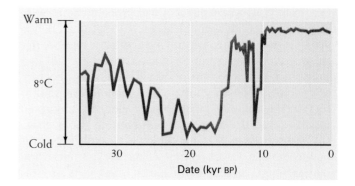

A temperature history of the surface waters of the North Atlantic ocean, based on the deuterium content of fossil shells.

and high density), and the cooler temperatures of the North Pacific limiting evaporation (thus promoting low salinity and low density).

The way in which the atmosphere and ocean are coupled in this process is important, but not at all well understood. In Broecker's discussion of some of the connections between ocean currents and climate, he emphasizes that Earth's climate does not respond to forcing in a smooth and gradual way, but in sharp jumps. Examples are provided by the average surface temperatures of the North Atlantic over the past 30,000 years, as deduced from the deuterium content of marine fossil shells. The record of that period, except for the most recent 10,000 years, is one of rapid fluctuations, with temperature changes of several degrees taking place in as little as a century. The concern this raises is that at some point the continuing increases in the concentrations of carbon dioxide and the other greenhouse gases may suddenly jolt the ocean–atmosphere system out of its current mode and into a drastically altered state that is hostile to the well-being of all or part of Earth's present population.

As Broecker points out, surprises, probably disagreeable ones, appear more likely than not for the ocean circulation–climate system. Climatologist J. W. C. White of the University of Colorado has commented forcefully on the anomalous stability of Earth's climate over the past 10,000 years:

We humans have built a remarkable socio-economic system during perhaps the only time when it could be built, when climate was stable enough to let us develop the agricultural infrastructure required to maintain an advanced society. We don't know why we have been so blessed, but even without human intervention, the climate system is capable of stunning variability. If the Earth had an operating manual, the chapter on climate might begin with a caveat that the system has been adjusted at the factory for optimum comfort, so don't touch the dials.

Biosphere–Atmosphere Coupling

One of the most controversial countermeasure ideas of all is the "Gaia hypothesis" developed by James Lovelock of Cornwall, England, who noted that the atmospheres of other planets in the solar system can be described perfectly by invoking chemistry and physics, but that Earth's atmospheric composition and history reflect as well the strong influence of biology. He suggests that microorganisms, plants, and animals act in such a way that Earth's environment becomes adjusted to states optimum for their maintenance. As formulated, his hypothesis requires not that the biosphere act consciously but that the adjustments arise from natural selection.

Lovelock explains his theory through a paradigm called "Daisyworld": a cloudless planet in which the environment is defined by a single variable—temperature—and the biota by a single species, daisies. Like most plant life, the daisies grow best over a restricted range of temperatures, the growth rate peaking near 23°C and falling to zero below 5°C and above 40°C, as shown in the figure on the following page. The daisies can be either dark or light; dark daisies will tend to absorb solar radiation and heat themselves and their surroundings, while light daisies will tend to reflect solar radiation and cool their surroundings. If average temperatures are below the optimum growth point 2, say at point 1 in the figure, daisies that absorb radiation better—the darker ones—will multiply faster. As they become abundant, the planetary temperature will rise in response to their absorption of radiation.

Should the temperature increase past point 2, however, the cooler white daisies will do better, and thus adjust the overall system in the direction of cooling. Provided there are no limitations of water and nutrients (as assumed in this simple system), the two types of daisies thus interact constructively with solar radiation to achieve equilibrium near 23°C. If the system were set going at point 3, above the optimum temperature, a similar argument would again hold. The assumption that growth is restricted to a narrow range of temperatures is crucial to the working of the mechanism, but all mainstream life is observed to be limited to this same narrow range; indeed, the peaked growth curve is common to other variables besides temperature, for example, pH levels and the abundance of nutrients.

The daisy theory can be applied not only to a situation with stable external forces, but to one with changing forces; a prime example is the increase in radiation produced by an aging star like the Sun. Under changing conditions, the Gaia model might yield a relatively uniform temperature over long periods of time, providing thus a new possible solution to the "weak Sun paradox," in which a fainter Sun than today's nonetheless generated an ancient world with surface oceans rather than global ice.

Lovelock and his colleagues have devised a variety of models more complex than the ones discussed above, some with a dozen color gradations of daisies, some with daisies, rabbits, and foxes. These systems turn out to be resilient even to such perturbations as abrupt loss of 40% of the daisies, as might occur in a bolide impact. Nonetheless, the systems are purposely designed as examples, and it is not claimed that they necessarily mimic the enormously complex and intertwined ecosystems on Earth.

The Gaia concept has inspired vigorous scientific discussion, partly because of the difficulty of finding a consistent criterion for defining what is meant by "optimum for the maintenance of living things." The optimum can clearly not apply to a single species in a group of species, because each is constantly involved in winning and losing relationships with others. Even taking Earth's biosphere as a whole, the idea of maintenance of an "optimum state" appears suspect. This quandary is illustrated by the dramatic counterexample to Gaia that occurred some 600 million years ago—a relatively recent time in the planet's history—when a rapid increase in atmospheric

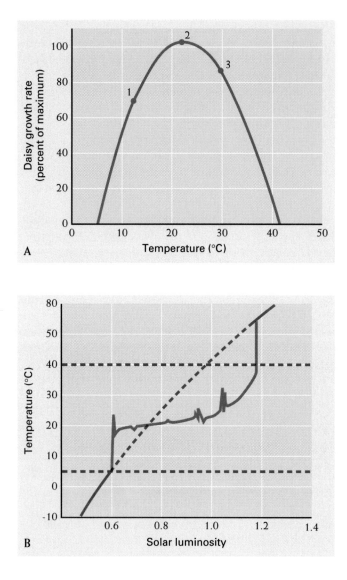

A

B

(Top) *Growth rates of daisies on Daisyworld as a function of tempera-ture, assuming adequate supplies of water and nutrients. The numbered points are discussed in the text.* (Bottom) *The evolution of temperature on Daisyworld as it is warmed by an evolving Sun, calculated by James Lovelock and colleagues. The red line is the result if life is not present, the green line if it is. The brief disruptions in the green line are produced by catastrophic death of 40% of the daisy population.*

nearly complete destruction of the biosphere from which it sprang. A more recent counter-Gaian example is the substantial decrease in atmospheric concentrations of the greenhouse gases carbon dioxide and methane that oc-curred during the last several ice ages. Rather than in-creasing so as to counteract the decrease in temperature, the biospherically related greenhouse gases declined in concentration and thus acted to augment the climate forcing that the variations in the planet's orbit had begun.

In response to such arguments, Lovelock has listed some possible current examples in which a positive medi-ating action of the biosphere appears to arise naturally from an action taken in local self-interest. For example, with Robert Charlson of the University of Washington, Lovelock and coworkers have noted that the production of the gas dimethyl sulfide (DMS) by planktonic algae in seawater can be seen as one such homeostatic feedback mechanism. Upon entering the atmosphere, the DMS is subject to oxidation by the hydroxyl radical, eventually to produce sulfuric acid (H_2SO_4):

$$DMS \xrightarrow{\text{several steps}} SO_2 \xrightarrow{\text{several steps}} H_2SO_4 \qquad (8.5)$$

Sulfuric acid, more stable as a liquid than a gas, rapidly gathers water molecules to form small, acidic droplets, and the droplets serve as efficient nuclei for the condensation of cloud drops. The more droplets, the whiter the clouds. Therefore, if the planktonic community is abundant and if sufficient DMS is transformed in this way, the result will be an enhancement in Earth albedo.

Because clouds scatter incoming solar radiation, the surface temperature and the level of photosynthesis under the clouds is expected to decrease, and thus the algal com-munity shrinks. As it does so, less DMS is released and the

oxygen proved lethal to most of the (anaerobic) life forms on Earth. The subsequent evolution of an almost entirely new biosphere (one that was adapted to the high oxygen supply and furthermore utilized solar energy much more efficiently than any living thing had done before) was "anti-Gaian" in the sense that it was predicated on the

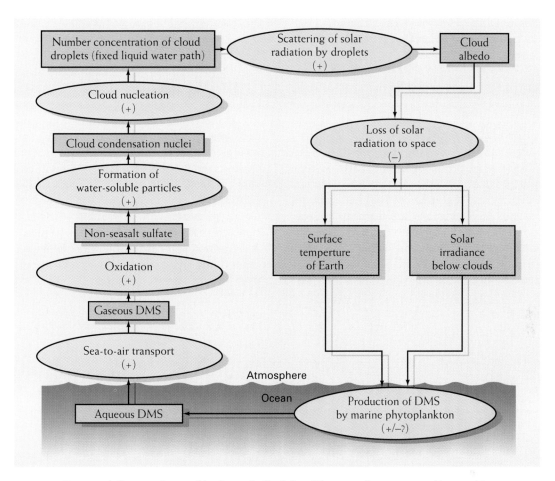

Conceptual diagram of a possible climate feedback loop. The rectangles are measurable quantities, and the ovals are processes linking the rectangles. The sign (+ or −) in the oval indicates the effect of a positive change of the quantity in the preceding rectangle on that in the succeeding rectangle. The most uncertain link in the loop is the effect of cloud albedo on DMS emission; its sign would have to be positive in order to regulate the climate.

cloud-forming potential of the atmosphere decreases, once more permitting solar radiation to penetrate to the surface and restore the algal community. The result is a naturally occurring regulation of the climate.

Charlson and Lovelock point out that this mechanism would tend to reduce atmospheric warming arising from increased concentrations of carbon dioxide, because the increased carbon dioxide leads to higher oceanic productivity and larger populations of DMS-emitting plankton, and thus to higher concentrations of cloud condensation nuclei. This results in more clouds and higher albedo—and therefore cooler water and climate. However, it appears from measurements of methane sulfonic acid in Antarctic ice cores (methane sulfonic acid, like sulfuric acid, is a DMS derivative) that oceanic DMS emissions were higher during glacial periods than during interglacial periods, contrary to what the Gaia hypothesis would propose.

Lovelock's concept was inspired by the demonstrable role of biology as a factor in atmospheric composition, yet that circumstance does not demonstrate biological control, advertent or inadvertent. It is clear, however, that the strong interactions of today's life forms with their environments are essential processes in the Earth system. Whether the coupling mechanisms interact as Gaia would predict, and whether the driving forces and the driven responders can readily be distinguished, remains to be seen. It is in this context that we once again consider the comparative characteristics of Earth and Venus. The total amount of carbon is similar on both planets, but on Venus the carbon is mostly present as atmospheric carbon diox-ide, thus strongly heating the planet, whereas on Earth it is mostly stored in biospheric reservoirs. In this manner, Earth's present biosphere clearly encourages the preservation of an environment suitable to its own sustainability.

Lovelock's ideas have raised the expectation in some minds that the planet has unexpected resources for resisting at least some of the perturbations caused by human development of the biosphere. However, any Gaia interactions operate on time scales that are much longer than those of human industrial development, and, in any case, will not necessarily function to benefit *Homo sapiens* in particular. The crucial question, then, is not whether Gaia-type systems are active, but whether humanity's actions can drive the Earth system beyond not only any short-term repair capabilities directed by humans, but also beyond any hypothetical Gaia repair capability. The answer to the crucial question depends on better under-

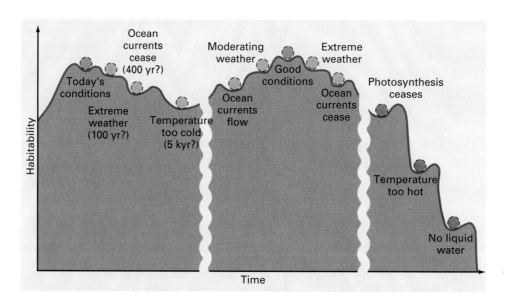

Earth's "hill of instability". Time increases from left to right, suitability for human habitation from bottom to top. A number of problematic metastable climate conditions are indicated. Transitory but temporary catastrophies due to such natural causes as major volcanic eruptions or bolide impacts are not shown, nor (of course) are metastable regions for which there is no previous known history, such as rapid warming due to anthropogenic greenhouse gas emissions.

standing of the interrelationships between biota and planet, an understanding that will take years to acquire. It is the great achievement of Lovelock and his collaborators that they have forced us to study Earth from this new perspective.

The Hill of Instability

A central message of this book is that, contrary to our experience and intuition, Earth's atmosphere and climate are not stable over time. The historical record, pieced together painstakingly from clues in rocks, ice, fossils, pollen, and other reservoirs of information, demonstrates that our present environment is a charmed one, subject to drastic change. Given this perspective, it is natural to identify the factors, discussed earlier in this book, that could upset our present relatively comfortable situation. Perhaps the least threatening factor is the predicted increase in the frequency of extreme weather events such as hurricanes and tornadoes. That circumstance would obviously be difficult for people living in the path of the storms, but would not materially disrupt global societal structures. More severe, however, would be a cessation of the flow of today's ocean currents, a prospect predicted by some global warming models. Major societal dislocation would result from such an occurrence. Still more severe would be the occurrence of a full glacial period, a happening very likely to take place in a few thousand years but probably not in the next few hundred.

Earth may go through many climatic cycles of increasing inhospitability and gradual recovery, so far as humanity and many other forms of life are concerned, and it may traverse as well more transitory climatic paths resulting from bolide impacts, human-related emissions, or other influences not yet identified. Eventually, however, the planet will gradually lose its three crucial life-enabling properties: enough carbon dioxide for photosynthesis, comfortable temperatures, and the availability of liquid water.

All of these environmental cycles may be pictured as an expanded version—a "hill of instability"—of the diagrams we presented in the first chapter. In this version, the near future is toward the left, the far future is toward the right, and depressions high on the hill represent environments conducive to life. Scientists know or can predict the existence of many of the regions of metastability on the hill, but are much less certain of the forces that precipitate a transition from one of those regions to another. They are also uncertain about whether Earth's history is a suitable guide to a future that has strong, new driving forces differing significantly from those of the past. To push our analogy to the limit, we wish to stay atop our hill of instability as long as we are able, rather than to be caught in a rockslide or a blind crevasse.

In this painting symbolizing the voyage of life, artist Thomas Cole creates as well an allegorical representation of all life on Earth as it attempts to navigate the rocks and shoals of unstable climate and environment.

Epilogue

As for the future, your task is not to foresee, but to enable it.
—Antoine de Saint Exupéry

Trends are not destiny.
—Rene Dubos

Clearly we humans are causing major changes in the composition of Earth's environments on every scale—local, regional, and global. Significant as these changes are, however, we have seen that they spring from a relatively few kinds of sources, and the sources causing the most serious damage are, in principle, almost wholly under human control: fossil fuel combustion, biomass burning, deforestation, agricultural practices, and industrial processes. One familiar with these issues would not ask, "Is global change occurring?" but rather, "Is global change occurring on such a scale or at such a rate that terrestrial life, including humankind, will find it difficult or impossible to adapt?" If we decide that the answer to the latter question is yes or even maybe, what then should we do about it?

A common response to uncertainty of many kinds is to "hedge one's bets" by making provisions that improve the odds of surviving low-probability catastrophes. For example, people wear seat belts as a safeguard against the occasional automobile accident and hide jewelry as a precaution against the occasional theft. They also buy insurance—against large medical expenses, fire damage to their homes, or early death. The insurance policy costs them a small fraction of their available income, and most people regard the expense as a necessary part of prudent financial management.

Almost no one really expects to have a house fire—or to see a major climate change—in his or her lifetime. Should a change in climate come to pass, however, the effects would be serious, perhaps catastrophic. Is it not then good management of humanity's resources to "buy some insurance" against rapid global climate change? The insurance need not be too expensive; many potentially advantageous actions are, in fact, quite inexpensive, and in some cases free. For instance, a "no regrets" approach to dealing with the possibility of global climate change might include such steps as improving vehicle fuel efficiency, revising building codes to conserve energy, upgrading public transportation, and switching from incandescent to fluorescent lighting. In the United States, the Office of Technology Assessment of the U.S. Congress estimates that these and other steps could be taken for less than 2% of the nation's total energy expenditures, well within reasonable insurance costs and with many steps paying for themselves.

It is difficult to produce many strong arguments against buying such a global insurance policy: the risks against which it mitigates are high, the cost modest. The main objection, of course, is that, as with all insurance policies, the costs are incurred immediately and the benefits, if any, will not be returned until an undetermined time in the future. Politically, such a situation is a recipe for inaction, since policymakers stand to bear the cost but not to enjoy the return on investment. A few perceptive politicians, however, are starting to demonstrate a new quality in their decision making: a sense of responsibility for coming generations as well as for their own, and a commitment to global as well as local well-being. These individuals recognize the extent to which change is becoming part of the daily lives of all the world's citizens, and they see that the political system must therefore transform itself from a vehicle dedicated to preserving stability to one dedicated to managing change. The transition will be enormously difficult and complicated, and those who strive to bring it about will need and deserve our full support.

What can Earth system science say to these farsighted servants of humanity about the probabilities of change in the atmosphere and climate? One way to summarize our perspective is by discussing the atmospheric and climatic stability on three time scales: near-term (a few decades), intermediate-term (a glacial to interglacial cycle, perhaps 20,000 years), and ultimate-term (the lifetime of the planet). When one does so, it is difficult to be optimistic about the environmental quality of the planet over the near term. On local and regional scales, there is abundant evidence that the resilience of natural systems is being overwhelmed—not in all locales, not for all measurement criteria, but more often than not. Only a return to the world of two centuries ago, when there was little combustion-generated power usage and little heavy industrial activity, would seem a certain way to mitigate this trend, given the present and predicted global population and current development scenarios, but such a transition is clearly unrealistic. Even could it be accomplished, stabilizing the atmosphere's CO_2 content at twice current levels would require global emissions to return below those of 1990.

Can the most gloomy prospects be avoided? Two unprecedented and seemingly unlikely occurrences would be required. The first is the invention (and quick and global implementation) of technologies designed to minimize the emissions of trace gases to the atmosphere. The second is a willingness in the more developed world to make the political and financial commitments necessary to bring these technologies to the less developed world without delay. These two achievements would demand global cooperation so extensive that they would be major surprises in their own right.

On the intermediate time scale, we can perhaps be somewhat more hopeful regarding the Earth system's stability—although the stability of that future time may have somewhat different characteristics than today's. The Earth has shown itself capable of enduring both catastrophic natural events and slower challenges such as the ice ages, events that had major effects on climate and biology, and of recovering to regenerate the amazing ecological

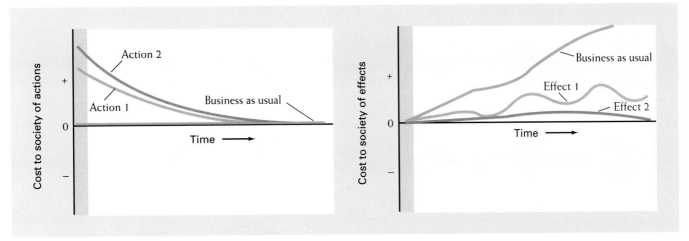

Conceptual diagrams of the (mostly immediate) costs to society of different actions taken to mitigate environmental damage and the (mostly eventual) costs to society of the effects of environmental damage. The different curves refer to different possible scenarios, the blue curves being "business as usual" (that is taking no mitigating actions at all), and the other curves suggesting two different levels of action. The light blue band on the left of each diagram indicates the period of influence of the decision makers who are now in a position to decide on appropriate actions. Economists are in the best position to suggest the form and magnitude of the costs of various actions; environmental scientists are in the best position to suggest the form and magnitude of the effects of various actions, if not their costs.

diversity we see today. Thus, even if we as humans are shortsighted enough to trigger a surprise whose consequences are as severe as a full glacial or thermal event, the planet may be able to regenerate itself in the intermediate term, although this outcome is in no way guaranteed. Instead, it is quite possible to imagine some runaway degenerative effect such as the release of large quantities of methane from the permafrost and methane hydrate sediments in polar regions, as a consequence of a long-lasting supergreenhouse warming.

On very long time scales, as we have seen, the planet will surely become incapable of sustaining life, as a consequence of increasing solar radiation. Whether Earth's ecosystem is terminated at that point and not before may or may not turn out to be under anyone's control. As a group, people are causing stresses to the biospheric system that are beyond all previous experience—and hoping that a surprise we cannot manage is not waiting in the wings. It is unlikely that scientists can be particularly useful for predicting the ultimate condition of the planet under such unprecedented stress, so prudence may be a better guide than science.

Does all of this lead inevitably to a dismal view of the future? The two of us think such a view unnecessary when applied to a planet we understand only poorly. Indeed, some great visionaries have foreseen exciting new eras in the life of our planet and of humanity on it, and we are reluctant to discount those possible outcomes. The optimistic viewpoint was given eloquent expression in 1902 by H. G. Wells:

It is possible to believe that all the past is but the beginning of a beginning, and that all that is and has been is but the twilight of the dawn. It is possible to believe that all the human mind has ever accomplished is but the dream before the awakening. We cannot see, there is no need for us to see, what this world will be like when the day has fully come. We are creatures of the twilight. But it is out of our race and lineage that minds will spring, that will reach back to us in our littleness to know us better than we know ourselves, and that will reach forward fearlessly to comprehend this future that defeats our eyes. All this world is heavy with the promise of greater things, and a day will come, one day in the unending succession of days, when beings, beings who are now latent in our thoughts and hidden in our loins, shall stand upon this earth as one stands upon a footstool, and shall laugh and reach out their hands amidst the stars.

Can we hope to realize Wells's vision? It will occur only if the planet, now peopled by the men and women whose descendants may bring that vision to pass, can be sustained until the great new dawn arrives. What, then, should be our actions as the current citizens of "Spaceship Earth"? One overriding guideline seems most appropriate: in seeking to develop Earth's resources for the benefit of all the people of today, we should seek at the same time to minimize in every way possible the attendant stresses on our planet. It is, indeed, the only one we have.

Further Reading

Chapter 1

T. E. Graedel, and P. J. Crutzen, *Atmospheric Change: An Earth System Perspective*, New York: W. H. Freeman and Company, 1993.
Most of the topics discussed in the present volume are treated in more detail in this college textbook.

T. J. Crowley, "The geologic record of climate change," *Reviews of Geophysics and Space Physics*, 21, 828–877, 1983.
A summary of the history of climate throughout geologic time.

Chapter 2

J. M. Wallace and P. V. Hobbs, *Atmospheric Science: An Introductory Survey*, San Diego: Academic Press, Inc., 1977.
This introductory textbook treats the atmospheric circulation in detail.

V. Ramanathan, B. R. Barkstrom, and E. F. Harrison, "Climate and the Earth's radiation budget," *Physics Today*, 42 (5), 22–32, 1989.
A brief article presenting the data from Earth satellites that quantify many of the processes in the radiation budget.

S. H. Schneider and R. Londer, *The Coevolution of Climate and Life*, San
Francisco: Sierra Club Books, 1989.
A popular but authoritative account of the science underlying weather and
climate.

W. S. Broecker and G. H. Denton, "The role of ocean–atmosphere
reorganizations in glacial cycles," *Geochimica et Cosmochimica Acta, 53*,
2465–2501, 1989.
A study of the salt transport system and its driving forces.

Chapter 3

B. Bolin and R. B. Cook, editors, *The Major Biogeochemical Cycles and Their
Interactions*, SCOPE 21, Chichester, U.K.: John Wiley and Sons, 1983.
Detailed descriptions of the cycling of sulfur, nitrogen, and phosphorus through
the Earth system.

J. A. Anderson, N. L. Hazen, B. E. McLaren, S. P. Rowe, C. M. Schiller, M. J.
Schwab, L. Solomon, E. E. Thompson, and E. M. Weinstock, "Free radicals
in the stratosphere: A new observational technique," *Science, 228*, 1309–
1311, 1985.
A short article describing an innovative scientific ballooning experiment and the
information derived from it.

P. Warneck, *Chemistry of the Natural Atmosphere*, San Diego: Academic Press,
Inc., 1988.
A scholarly volume that treats the processes of atmospheric chemistry in detail
and with rigor.

B. J. Finlayson-Pitts and J. N. Pitts, Jr., *Atmospheric Chemistry: Fundamentals and
Experimental Techniques*, New York: John Wiley & Sons, 1986.
A textbook that provides clear explanations and extensive bibliography on both
theoretical and experimental approaches to atmospheric chemistry.

Chapter 4

T. J. Crowley and G. R. North, *Paleoclimatology*, Oxford: Oxford University Press,
Inc., 1991.
An authoritative summary on observational and theoretical approaches to the
climates of the past.

S. M. Stanley, *Earth and Life Through Time*, 2d edition, New York: W. H. Freeman and Company, 1989.
An introduction to geophysics and paleontology, written from the standpoint that the physical and biological histories of Earth are inextricably intertwined.

R. J. Delmas, "Environmental information from ice cores," *Reviews of Geophysics*, 30, 1–21, 1992.
This review describes the formation of polar ice, its entrapment of atmospheric gases and particles, and the historical records of climate and chemistry that emerge from ice core analyses.

R. S. Bradley, editor, *Global Changes of the Past*, Boulder, Colo.: University Corporation for Atmospheric Research, 1991. (Available from UCAR, P.O. Box 3000, Boulder, CO 80307-3000.)
This multi-authored volume from a summer workshop is the best comprehensive summary of the various methods used to deduce Earth history from samples of the past.

Chapter 5

Committee on Tropospheric Ozone Formation and Measurement, *Rethinking the Ozone Problem in Urban and Regional Air Pollution*, Washington, D. C.: National Academy Press, 1991.
A comprehensive summary of data, theory, and policy relating to the formation of urban ozone, the most important component of photochemical smog.

P. M. Irving, editor, *Acidic Deposition: State of Science and Technology*, Washington, D.C.: U.S. Govt. Printing Office, 1991.
The summary report of the first decade of the U.S. National Acid Precipitation Assessment Program.

O. B. Toon and R. P. Turco, "Polar stratospheric clouds and ozone depletion," *Scientific American*, 264, (6), 68–74, 1991.
A readable introduction to the interaction of gases and cloud particles in the polar stratosphere.

U. S. Environmental Protection Agency, *National Air Pollutant Emission Trends, 1900–1992*, Research Triangle Park, N.C.: EPA-454/R-93-032, 1993.
A history of United States atmospheric emissions in the twentieth century. Some global data are also included.

Chapter 6

J. T. Houghton, G. J. Jenkins, and J. J. Ephraums, editors, *Climate Change: The IPCC Scientific Assessment*, Cambridge, U.K.: Cambridge University Press, 1990.
This volume, produced as background information for the 1992 United Nations Conference on Environment and Development, is the consensus of 170 scientists worldwide on the status of the global warming theory.

Single topic dedicated issue: "Managing Planet Earth," *Scientific American, 261* (3), 1989. (Also available in paperback from W. H. Freeman and Company, New York.)
A brief, readable look at the environmental hazards facing the planet and how the world community can work together to improve our common future.

B. L. Turner, II, W. C. Clark, R. W. Kates, J. F. Richards, J. T. Matthews, and W. B. Meyer, *The Earth as Transformed by Human Action*, Cambridge, U.K.: Cambridge University Press, 1990.
An extensive review of human-produced global and regional changes in the biosphere over the past 300 years.

W. J. Kaufman and L. L. Smarr, *Supercomputing and the Transformation of Science*, New York: Scientific American Library, 1993.
A book, in the same series as the present one, which explores the capabilities and limits of computation and presents examples from many scientific specialties.

D. Ojima, editor, *Modeling the Earth System*, Boulder, Colo.: University Corporation for Atmospheric Research, 1992.
This multi-authored volume from a summer workshop is the best comprehensive summary of Earth system models. (Available from UCAR, P.O. Box 3000, Boulder, CO 80307-3000.)

Chapter 7

A. Berger, "Milankovich theory and climate," *Reviews of Geophysics, 26,* 624–657, 1988.
A technical description of the variations of Earth in its orbit about the Sun, and the history and prospects of variations in the orbital parameters over time.

T. E. Graedel, I.-J. Sackmann, and A. I. Boothroyd, "Early solar mass loss: A potential solution to the weak Sun paradox," *Geophysical Research Letters, 18,* 1881–1884, 1991.

This brief paper proposes an explanation for how a faint Sun in early Earth's lifetime could have maintained planetary temperatures above the freezing point of water.

T. J. Ahrens and A. W. Harris, "Deflection and fragmentation of near-Earth asteroids," *Nature, 360,* 429–433, 1992.
A short discussion of the risk to Earth of asteroid impact and the possibility of actively intercepting asteroids on Earth-crossing paths.

Chapter 8

J. W. Tester, D. O. Wood, and N. A. Ferrari, editors, *Energy and the Environment in the 21st Century,* Cambridge, Mass.: MIT Press, 1991.
This volume provides a comprehensive reference to world energy requirements for the next century and of environmentally acceptable means of meeting them.

R. M. White, "The great climate debate," *Scientific American, 263,* (1), 36–43, 1990.
A concise summary of the arguments for, about, and against the theory of global warming.

S. H. Schneider and P. J. Boston, editors, *Scientists on Gaia,* Cambridge, Mass.: MIT Press, 1992.
Forty-four authors discuss the foundations of the Gaia theory, the mechanisms by which it might occur, and the public policy implications that result.

F. S. Rowland and M. J. Molina, "Ozone depletion: 20 years after the alarm," *Chemical and Engineering News, 72* (33), 8–13, 1994.
The originators of the explanation for the impact of chlorofluorocarbons on stratospheric ozone discuss experimental evidence supporting that proposal, as well as the international policy actions that have resulted from it.

Epilogue

World Commission on Environmental Development, *Our Common Future,* Oxford-New York: Oxford University Press, 1987.
An examination of the critical environment and development problems of the planet, and of realistic proposals to mitigate their effects.

W. C. Clark and R. E. Munn, editors, *Sustainable Development of the Biosphere*, Cambridge, U.K.: Cambridge University Press, 1986.
A multi-authored exploration of human social and technical development, changes in the world environment, and workable responses to those pressures.

Sources of Illustrations

Illustrations by Network Graphics, Gregory Wakabayashi,
Fine Line Illustrations, and Tomo Narashima

Cover Image
Pieter, the Younger, Bruegel. *Winter Landscape*. 1601.
Kunsthistorisches Museum, Vienna, Austria. Erich
Lessing/Art Resource

Frontispiece
NASA

Facing page 1
David Whillas, CSIRO Atmospheric Research,
Australia

page 2
Peter Ginter/Bilderberg

page 5
NASA

page 6
Colin Raw/Tony Stone Images

page 10
Jon Bradley/Tony Stone Images

page 13
Hank Morgan/SPL, Photo Researchers

page 17
Adapted from S. H. Schneider and R. Londer, *The
Coevolution of Climate and Life*. Copyright 1989 by
Schneider and Londer, Sierra Club Books, San
Francisco

page 22
Sandro Botticelli. *The Birth of Venus*. Uffizi, Florence.
Scala/Art Resource

page 23
P. J. Wyllie, *The Dynamic Earth: Textbook in Geosciences*.
Copyright 1971 by John Wiley & Sons

page 25
John Turner/Tony Stone Images

page 27
Adapted from J. W. M. la Rivière, *Scientific American*,
261 (3), 80–94. Copyright 1989 by Scientific American,
Inc.

page 28
Collection of the New York Historical Society

page 29
Earth Satellite Corp.

page 31
Adapted from J. M. Wallace and P. V. Hobbs, *Atmospheric Science: An Introductory Survey. Copyright 1977
by Academic Press, Inc.*

page 32
Ted Spiegel/Black Star

page 33
Joseph Waters, Jet Propulsion Laboratory

page 34
SPL/Photo Researchers

page 36
Copyright 1987, The New Yorker Magazine, Inc.

page 39
(*upper left*) NASA

page 39
(*upper right and lower right*) Adapted from J. A. Anderson, N. L. Hazen, B. E. McLaren, S. P. Rowe, C. M. Schiller, M. J. Schwab, L. Solomon, E. E. Thompson, and E. M. Weinstock, Free radicals in the stratosphere: a new observational technique, *Science, 228*, 1309–1311, 1985; (*lower left*) W. H. Brune, E. M. Weinstock, S. J. Schwab, R. M. Stimpfle, and J. G. Anderson, Stratospheric CIO: In-situ detection with a new approach, *Geophysical Research Letters, 12*, 441–444, 1985

page 40
NASA, Goddard Space Flight Center

page 41
NASA

page 44
South Coast Air Quality Management District

page 45
S. A. Changnon, Jr., Rainfall changes in summer caused by St. Louis, *Science, 205*, 402–404. Copyright 1979 by AAAS

page 46
Adapted from W. S. Cleveland, B. Kleiner, J. E. McRae, and J. L. Warner, Photochemical air pollution: Transport from the New York City area into Connecticut and Massachusetts, *Science, 191*, 179–181. Copyright 1976 by AAAS

page 47
(*right*) H. B. Singh, W. Viezee, and L. J. Salas, Measurements of selected C_2–C_5 hydrocarbons in the troposphere: Latitudinal, vertical, and temporal variations, *Journal of Geophysical Research 93*, 15, 861, 1988 (*left*) L. P. Steele, et al., The global distribution of methane in the troposphere, *Journal of Atmospheric Chemistry, 5*, 125–171. Copyright 1987 by Kluwer Academic Publishers

page 48
A. R. Ravishankara, NOAA, Aeronomy Laboratory

page 51
(*bottom*) Will McIntyre/Photo Researchers

page 52
Courtesy of National Atmospheric Deposition Program, Colorado State University, Fort Collins

page 53
Westfälisches Amt für Denkmalpflege

page 54
Tony Stone Images

page 55
C. M. Benkovitz, Environmental Chemistry Division, Brookhaven National Laboratory

page 58
Andrew Knoll, Harvard University

page 62
Spence Titley

page 63
R. P. Wayne, *Chemistry of Atmospheres*, 2d Ed. Copyright 1991 by Clarendon Press, Oxford, U.K.

page 66
T. M. L. Wigley, Climate and paleoclimate: what can we learn about solar luminosity variations?, *Solar Physics, 74*, 435–471. Copyright 1981 by Kluwer Academic Publications

page 67
T. E. Graedel, I.-J. Sackmann, and A. I. Boothroyd, Early solar mass loss: A potential solution to the weak

sun paradox, *Geophysical Research Letters*, 18, 1881–1884, 1991

page 68
(*top left*) Israel Geological Survey; (*bottom right*) Chase Studio, Inc. Photography by Dan Rockafellow

page 69
Deep Sea Drilling Project

page 70
Adapted from K. G. Miller, R. G. Fairbanks, and G. S. Mountain, Tertiary oxygen isotope synthesis, sea level history, and continental margin erosion, *Paleoceanography*, 2, 1–19, 1987

page 71
Adapted from K. Winn and C. R. Seotese, *Phanerozoic Paleographic Maps*, University of Texas Institute of Geophysics, Tech. Rpt. 84, 31 pp., 1987

page 72
(*top left*) Woods Hole Oceanographic Institution; (*top right*) Chevron Corp.; (*bottom*) Scientific American, January 1990; (*left*) Bruce Corner, Lamont-Doherty Geological Observatory of Columbia University (*center and right*) Dee L. Breger, Lamont-Doherty Geological Observatory of Columbia University

page 73
Adapted from A. Hallam, Pre-Quaternary sea-level changes, *Annual Reviews of Earth and Planetary Science*, 12, 205–243. Copyright 1984 by Annual Reviews, Inc.

page 74
Space Telescope Institute/NASA

page 75
Adapted from L. W. Alvarez, Mass extinctions caused by large bolide impacts, *Physics Today*, 40, (7) 24–33, 1987

page 77
(*top*) A. Berger, Milankovich theory and climate, *Reviews of Geophysics*, 26, 624–657, 1988 (*bottom*) Adapted from J. Imbrie and J. Z. Imbrie, Modeling the

climatic response to orbital variations, *Science*, 207, 943–952. Copyright 1980 by the AAAS

page 78
Tom Bean

page 79
R. J. Delmas, laboratorie de glaciologie et géophysique de l'environnement, Centre National de la Recherchf Scientifique

page 80
R. J. Delmas, Environmental information from ice cores, *Reviews of Geophysics*, 30, 1–21, 1992

page 82
Adapted from J. H. McAndrews, in *Quaternary Paleoecology*, E. J. Cushing and H. E. Wright, Jr., eds., pp. 218–236, University of Minnesota Press, 1967

page 83
Pieter, the Younger, Bruegel. *Winter Landscape.* 1601. Kunsthistorisches Museum, Vienna, Austria. Erich Lessing/Art Resource

page 84–85
Adapted from COHMAP Members, Climatic changes of the last 18,000 years: Observations and model simulations, *Science*, 241, 1043–1052. Copyright 1988 by the AAAS

page 86
Adapted from T. Webb III, J. Kutzbach, and F. A. Street-Perrott, in *Global Change*, T. Malone and J. Roederer, eds., pp. 182–218, Cambridge University Press, U.K. 1985

page 88
Paul Crutzen, National Center for Atmospheric Research, University Corporation for Atmospheric Research, National Science Foundation

page 92
Adapted from R. J. Delmas, Environmental information from ice cores, *Reviews of Geophysics*, 30, 1–21, 1992

page 93
Shawn Henry/SABA

page 94
C. U. Hammer, H. B. Clausen, and W. Dansgaard, Greenland ice sheet evidence of post-glacial volcanism and its climatic impact, *Nature, 288*, 230–235. Copyright 1980 by Macmillan Magazines Limited

page 96
Constantin Meunier, *In the Black Country*. Musee d'Orsay, Paris. Erich Lessing/Art Resource

page 97
(*top*) H. Friedli, H. Lötscher, H. Oeschger, U. Siegenthaler, and B. Stauffer, Ice core record of the $^{13}C/^{12}C$ ratio of atmospheric CO_2 in the past two centuries, *Nature, 324*, 237–238. Copyright 1986 by Macmillan Magazines Limited; (*bottom*) Donald Cahoon, NASA Langley Research Center

page 98
(*top*) David Madison/Tony Stone Images (*bottom*) M. A. K. Khalil and R. A. Rasmussen, Atmospheric methane: Trends over the last 10,000 years, *Atmospheric Environment, 21*, 2445–2452. Copyright 1987 by Pergamon Press, plc.

page 99
Angus M. Mackillop/Tony Stone Images

page 100
Norwegian Meteorological Institute, Oslo

page 101
(*left*) Ed Pritchard/Tony Stone Images

page 102
A. Volz and D. Kley, Evaluation of the Montsouris series of ozone measurements made in the nineteenth century, *Nature, 332*, 240–242. Copyright 1988 by Macmillan Magazines Limited

page 103
Hank Morgan/Photo Researchers

page 104
Environmental Protection Agency, *National Air Quality and Emission Trends Report*, EPA-450/4-84-029, Research Triangle Park, N.C. 1985

page 105
Courtesy of C. D. Keeling, Scripps Institution of Oceanography, La Jolla, Ca.

page 106
Adapted and updated from J. C. Farman, B. G. Gardiner, and J. D. Shanklin, Large losses of total ozone in Antarctica reveal seasonal interaction, *Nature, 315*, 207–210, 1985

page 107
(*top*) A. J. Krueger, Goddard Space Flight Center, NASA; (*bottom*) Copyright 1992 The New Yorker Magazine, Inc.

page 109
(*bottom*) J. E. Frederick and A. D. Alberts, Prolonged enhancement in surface ultraviolet radiation during the Antarctic spring of 1990, *Geophysical Research Letters, 18*, 1869–1871, 1991; (*inset*) A. J. Krueger, Goddard Space Flight Center, NASA

page 110
(*top*) Courtesy of The World Meteorological Organization, Geneva; (*bottom*) J. W. Elkins, et al., Decrease in the growth rates of atmospheric chlorofluorocarbons 11 and 12, *Nature, 364*, 780–783. Copyright 1993 by Macmillan Magazines Limited

page 112
James P. Blair, Copyright © National Geographic

page 114
John Dunlop, Woods Hole Oceanographic Institution

page 118
David Woodfall/Tony Stone Images

page 120
L. E. Manzer, The CFC-ozone issue: Progress on the

development of alternatives to the CFCs, *Science, 249,* 31–35. Copyright 1990 by AAAS

page 121
J. T. Houghton, G. J. Jenkins, and J. J. Ephraums, eds., *Climate Change: The IPCC Scientific Assessment.* Copyright 1990 by Cambridge University Press

page 122
J. T. Houghton, G. J. Jenkins, and J. J. Ephraums, eds., *Climate Change: The IPCC Scientific Assessment.* Copyright 1990 by Cambridge University Press

page 125
Donald Wuebbles, Lawrence Livermore Laboratories

page 126
Malcolm Ko, Atmospheric and Environmental Research, Inc., Cambridge, Mass.

page 130
Paul Crutzen

page 131
A. G. Russell and G. M. McRae, Carnegie Mellon University; Michael McNeill, William Sherman, Mark Bajuk, NCSA

page 132
S. Manabe and R. J. Stouffer, Sensitivity of a global climate model to an increase of CO_2 concentration in the atmosphere, *Journal of Geophysical Research, 85,* 5529–5554, 1980

page 133
(*top*) David C. Turnley/Detroit Free Press; (*bottom*) Library of Congress

page 134
David Rind/NASA Goddard Institute for Space Studies

page 136
Adapted from Committee on Global Change, *Toward an Understanding of Global Change,* National Academy Press, Washington, D.C., pp. 72–73, 1988

page 138
R. J. Charlson, J. Langner, H. Rodhe, C. B. Leovy, and S. G. Warren, Perturbation of the northern hemisphere radiative balance by backscattering from anthropogenic sulfate aerosols, *Tellus,* 43AB, 152–163. Copyright 1991 by Munksgaard International Publishers Ltd.

page 139
Albert J. Semter, Naval Postgraduate School, and Robert M. Chervin, National Center for Atmospheric Research. Image generated by Randy Cubrilovic

page 140
Courtesy of S. Manabe, NOAA Geophysical Fluid Dynamics Laboratory, Princeton, N.J.

page 142
Virgil L. Sharpton, Lunar and Planetary Institute, Houston

page 144
(*left*) Adapted from W. Broecker, *How to Build a Habitable Planet,* Eldigio Press, Palisades, N.Y., 1987

page 144
(*right*) A. Berger, Milankovich theory and climate, *Reviews of Geophysics,* 26, 624–657, 1988

page 145
Adapted from H. H. Lamb, *Climate: Present, Past, and Future,* Vol. 2. Copyright 1977 by Methuen & Co.

page 146
Adapted from W. L. Prell and J. E. Kutzbach, Monsoon variability over the past 150,000 years, *Journal of Geophysical Research,* 92, 8411–8425, 1987

page 147
(*left*) Constructed from information in A. Berger, The Earth's future climate at the astronomical time scale, in *Future Climate Change and Radioactive Waste Disposal,* C. Goodess and S. Pakstikof, eds., Norwich, U.K., 1989; (*right*) Lonnie G. Thompson, Byrd Polar Research Institute, Ohio State University

pages 148–149
Adapted from K. Winn and C. R. Scotese, *Phanerozoic Paleographic Maps*, University of Texas Institute of Geophysics, Tech. Rpt. 84, 31 pp., 1987

page 150
Courtesy of I.-J. Sackmann, California Institute of Technology, Pasadena, Ca

page 152
(*top*) K. Caldiera and J. F. Kasting, The life span of the biosphere revisited, *Nature, 360,* 721–723. Copyright 1992 by Macmillan Magazines Limited; (*bottom*) from J. B. Kaler, *Stars,* Scientific American Library

page 154
Regis Lefebure/Third Coast Stock

page 158
J. B. Kumar, Lockheed

page 159
(*top*) Adapted from C. P. Rinsland, J. Levine, A. Goldman, N. D. Sze, M. K. W. Ko, and D. W. Johnson, infrared measurements of HF and HCl total column abundances above Kitt Peak, 1977–1990: Seasonal cycles, long-term increases, and comparisons with model calculations, *J. Geophysical Research, 96,* 15523–15540, 1991; (*bottom*) J. W. Walters, L. Froidevaux, W. G. Read, G. L. Manney, L. S. Elson, D. A. Flower, R. F. Jarnot, and R. S. Harwood, Stratospheric ClO and ozone from the Microwave Limb Sounder on the Upper Atmospheric Research Satellite, *Nature 362,* 597–602

page 161
Star Tribune

page 163
R. J. Cicerone, S. Elliott, and R. P. Turco, Reduced Antarctic ozone depletions in a model with hydrocarbon injections, *Science, 254,* 1191–1193. Copyright 1991 by the AAAS

page 166
Adapted from W. Broecker, Unpleasant surprises in the greenhouse?, *Nature, 328,* 123. Copyright 1987 by Macmillan Magazines Limited

page 168
Adapted from J. E. Lovelock, Geophysiology: A new look at Earth science, *Bull Am. Met. Soc., 67,* 392–397, 1986

page 169
R. J. Charlson, J. E. Lovelock, M. O. Andreae, and S. G. Warren, Oceanic phytoplankton, atmospheric sulphur, cloud albedo and climate, *Nature, 326,* 655–661. Copyright 1987 by Macmillan Magazines Limited

page 172
Thomas Cole. *The Voyage of Life: Manhood.* Ailsa Mellon Bruce Fund, © 1994 Board of Trustees, National Gallery of Art, Washington, 1842, oil on canvas

page 175
Adapted from a diagram devised by M. McFarland, E. J. DuPont de Nemours, Wilmington, De.

INDEX

Mt. Everest, 5
Mt. Mitchell, North Carolina, 51
Mt. Pinatubo, 5, 32–33, 93
Mylona, Sophia, 99

National Center for Atmospheric
    Research, 13, 88
Natural gas combustion, as source of
    CO_2, 95–96
New York City, "Blizzard of '88," 28
Nitrate, historical concentrations of,
    92
Nitric acid
    as reservoir molecule, 38
    dissolution of, 50
Nitrogen dioxide
    smog chemistry, 42–43
    stratospheric concentrations, 41
Nitrous oxide
    as greenhouse gas, 15
    atmospheric budget imbalance, 111
    chemistry, 37
    global warming potential (GWP),
        120–121
North America
    predicted climate, 135
    water flow, 26

Oceans
    deep water currents, 23
    surface currents, 23
    temperature history, 166
    volume, 23
Office of Technology Assessment,
    174
Ohio River Valley, 52
One-dimensional (1D) model,
    117–118

Oude, Rijn, 8
Oxides of nitrogen
    as catalyst for ozone destruction,
        37, 42–43
    as catalyst for ozone production,
        43
Oxidizers, atmospheric, 127
Oxygen, 12
    atmospheric budget imbalance, 111
    evolutionary history, 62–63
Oxygen isotopes, as temperature indi-
    cators, 67, 77
Ozone
    Antarctic concentrations, 106–107
    and protection, 107
    and unpredicted influences, 36
    as atmospheric scavenger, 42
    column abundances, 31
    evolutionary history, 62–63
    formation in stratosphere, 12
    historical concentrations, 102
    surface concentrations, 129–130
    transport of, 46
Ozone depletion potential (ODP),
    119–120
Ozone depletion, countermeasures,
    162–163
Ozone hole, 106–107

Paleoclimatology, 64
Paleozoic, carbon dioxide concentra-
    tions, 91
Pangaea, 7, 71, 75
Paris Observatory, 102
Particulate matter, urban, 101
Pedosphere, 4
Petroleum combustion, as source of
    CO_2, 95–96
pH, 50

Other Books in the Scientific American Library Series